LIFE. CAMERA. ACTION.

HOW TO TURN YOUR MOBILE DEVICE
INTO A FILMMAKING POWERHOUSE

LIFE
CAMERA
ACTION

DAVID BASULTO

LIFE. CAMERA. ACTION.
*How to Turn Your Mobile Device
into a Filmmaking Powerhouse*

ISBN   978-1-61961-568-7   *Paperback*
       978-1-61961-569-4   *Ebook*

LIONCREST
PUBLISHING

To my rock, and partner in life—my wife, Loren. Without your love and dedication, none of this would have happened. You've given me the two greatest joys in my life: our relationship, and our son, Alexander. Your belief in me led to the third, iOgrapher.

To Alexander: Every day, you make me a better father and man. Watching your creativity and boundless potential unfold is a thrill. I can't wait to see what you become.

To Shauna Soltis: You gave me the opportunity of a lifetime—my teaching job at San Marino High School. You made a great job even better, and without teaching, I surely wouldn't have written this book.

To my business partner and the brother I wish I'd had, Tony Basalari: None of this would exist without your son's presence in my classroom, your investment, and your dedication to the company. Let's keep growing and reach the stars one day, so we can retire and bask in our glory.

To Sally DeWitt: Your investment as an iOgrapher investor has been invaluable. Thank you for your support and believing in me, iOgrapher, and the future of all that we do!

To Shirley Bozek: Your early donation in our first Kickstarter campaign helped launch the iOgrapher product line. Thank you for your support from the beginning!

To all of my students, whose creativity inspired my invention of the iOgrapher: I'll always love you, and I'll be watching as you shape this wonderful planet.

To you, readers and iOgrapher supporters: Without you, I'm nothing.

CONTENTS

||||||||||||||||||||||||||||||

INTRODUCTION

I grew up in what I call the Dark Age of Creativity. As a ten-year-old who wanted to make movies with my friends, I would have needed enough money to buy a Super 8 or 16-millimeter camera, an editing device like a Moviola, and a projector. I would have then needed the time and effort to develop film and gather all my friends in one place. It would have cost thousands of dollars and hundreds of hours, plus required expertise that kids don't yet have. Because today's tools weren't accessible to me, I didn't get to develop the creative side of my brain. It was a shame.

These days, a ten-year-old can make professional-quality films—and become a journalist, vlogger, teacher, and much more—with just his or her iPhone or iPad and a few free apps. The dark ages are over, and the possibilities are *endless.*

As a teacher watching my kids blossom in this new world of mobile filmmaking, I was inspired, and so I became a first-time entrepreneur at fifty-one years old. Decades later, the creative side of my brain has been unleashed, and I want to help kids and adults to the same. The tools finally exist for anyone, anywhere, to tell any story he or she wants.

As a kid, I was a standout football player. I ended up at the Oregon State University on a football scholarship, but my Latina mother couldn't stand the distance, before Skype and FaceTime. I quit, played in junior college, then moved on to Fresno State, which was closer to home. The coaches said I could make the NFL if I switched from defensive end to nose guard, but that meant I'd play over the ball and get my skull crushed on every snap. Playing in the NFL was my dream, but daily concussions weren't. I refused, and I transferred to San Diego State, where I could play defensive end while studying business. My NFL dream died, but I got recruited by Dean Witter Reynolds, a big stockbroker. My next dream was to be like the guys in the movie *Wall Street*, and that had come true.

I worked on the sixty-third floor of the old World Trade Center and lived it up as a young bachelor in New York

City. I was out with friends one night, one of whom was a model, and a lady approached. She was a big casting director, and she saw long-haired, burly me and told me I had a "great look." I was still in shape from football. She asked if I wanted to get into acting, as an extra in *Rocky V*.

I got the part, met Sylvester Stallone and the cast, and fell in love with Hollywood. I remembered my stunted creative side, and realized my analytical, mathematical brain wanted company. I quit Wall Street and moved back to LA. My parents freaked out as I went from a twenty-three-year old stockbroker making $80,000 a year to an extra making forty dollars a day.

Six months later, I earned my Screen Actors Guild (SAG) card and played a mafia guy in a Steven Seagal movie, when he was in his prime. I began to do feature films and television, working on *Cheers*, *Mad About You*, and *NewsRadio*. My biggest role was playing opposite Edward Norton in *American History X*. Acting life was fulfilling, but paychecks were irregular and stressful, and I needed odd jobs on the side to pay bills.

One day I was driving from Pasadena to an audition in Santa Monica. This was a drive that should have taken fifteen minutes, but LA traffic made it take an hour. I thought about my auditions. I'd make this long, stressful drive, pay for gas, pray to get a job...and then my audition would consist of saying two words. I'd do all of this work for a small chance to get a bit part, and I couldn't take it anymore. I always saw movie producers sitting on sets, smoking cigars, and telling people what to do. I decided I'd become one, because producing was more comfortable than acting.

I already knew the film industry and money side, so my first movie took just forty-five days to create. I raised 2.5 million dollars and made a mafia film called *18 Shades of Dust*, starring Academy-Award nominee Danny Aiello. Much of the cast wound up being on *The Sopranos*.

Producing was a blast; I produced five more movies and eventually worked with Mel Gibson's Icon Entertainment, then became an executive producer at Lifetime Television.

After getting married, the eighteen hours on set, plus constant travel, got to me. I was exhausted, and I needed to find my next journey. I "retired" and spent time with our beautiful, newborn son, and made movies on my own. That's when I started to get into technology. Panasonic had released a camera called the DVX100, a little consumer camera that could shoot with exceptional film quality, twenty-four frames per second. I couldn't believe what I could do with that tiny camera.

I made a small feature film for $10,000 that took ten days to shoot and sold it to Blockbuster. Then I made a documentary about my wife, and it won film festivals. We edited it together on my computer with Adobe software. I was amazed at how fun and simple it all was, and I was thrilled I could finally be a storyteller.

I became an evangelist for different software and started a blog called *Filmmaking Central*. It still gets a lot of traffic. I was also one of the early podcasters. I couldn't believe that people wanted to listen as I spoke into a microphone and told stories about the film industry.

One day, I got a Facebook message from a high-school friend who was working at our rival high school. She offered me a job teaching media and animation. I interviewed and fell in love with the energy and curiosity of the kids. I couldn't wait to show them what they could create with modern technology; I couldn't wait to help them do what I couldn't as a kid. I got my teaching credential and started my new life. Teaching was the most rewarding job I'd ever had—except for the financial aspect, which is a tragedy for another book.

As I taught, I tinkered. Our budget left us with five cameras for 130 students. They were quality DSLRs and broadcast-type cameras, but I couldn't loan them out on weekends, because I needed them to film football games for boosters. I started to let my class use iPhones and iPads. They got much more work done, but the footage was shaky, the audio was horrible, and so was the lighting.

Filming with iPhones and iPads had potential, but we needed to fix those problems. I tinkered with the idea of a case with handles on both sides, to fix our main issue of stability. Nothing on the market existed to solve this problem, so I made a prototype myself, using a 3D-printer in my classroom.

I was amazed; this wasn't the dark ages I'd grown up in. Technology was evolving so fast. I could fit an iPad Mini in my (admittedly defensive-end-sized) pockets. *Anyone* with the desire to make a movie could make a movie. They could also record events, learn stop-motion animation, become a photographer, and do basic editing. The world had opened for storytellers of all ages, experience levels, and budgets.

Then, one of my kids introduced me to Kickstarter. We raised money and got covered by *Forbes*, *Mashable*, the *Wall Street Journal*, and other publications; they loved the story of a teacher unleashing his students' creativity by using iPads. We reached our financial goal, but then I realized I had no idea how to actually produce the things.

Again, my students saved me. One of their moms owned an injection modeling company. We met and decided to make everything locally, in California, and she got the ball rolling. Another student's parents owned a packaging company.

Another student's parent, a senior executive at Disney with whom I'd become close friends, took me and the students to Disney Studios. I brought my iPad and handled case, which I called an iOgrapher, and he was struck. "What's that?" He wrote a big check, and all of a sudden, I was a first-time entrepreneur at fifty-one years old.

I juggled teaching and entrepreneurship for a year, until we signed a deal with Best Buy, to enter one thousand stores. If I wanted to get serious about iOgrapher, I needed to quit teaching. I didn't have the time to fully devote myself to the kids. That was near the end of 2015. I've been a full-time entrepreneur ever since.

To the aspiring NFL player back in college, my meandering life might have seemed like a failure. But all of those "failures" led me exactly where I wanted to be, as an entrepreneur helping people of all backgrounds to tell stories. My students had inspired me to change my goals, through their curiosity and creativity. I'm currently the happiest and most successful I've ever been. I'm an entrepreneur creating things I love, and I still have time to spend with my wife and son.

When the second iPad came out (the first one equipped with a camera), I was amazed. I could make little videos, and I was able to teach my two-year-old son basic stop-motion animation with his Hot Wheels toys. He *got* it. It was intuitive. We started making fun things together.

Students are often bound by rigid curricula and a lack of tools, so I let them play with the intuitive devices already in their pockets. Most had iPhones or Samsungs, and I had a few iPads. Their curiosity led as they tapped and played and created, and they learned at an incredible rate. iOS devices cost around $500 each, but they're hard to break, and they're cheaper and more intuitive than a $2,000 DSLR or $3,000 broadcast cameras. I could trust my students and allow them to play and express themselves. They *got* it. It was clearly the future of education.

iOS devices in particular are intuitive; you pick it up and say, *Let me press this.* You discover as you go. I'd show the students a couple apps, and they'd be off to the races. Toward the end of my tenure, we were streaming football games live, with four different angles, to YouTube. We were shooting the school play, and filming full-length movies. Operating video equipment no longer required expertise, meaning children of all ages could produce quality films.

Teachers started getting into it. They filmed Shakespeare recitals in English class, and then played the film back to students so they could correct mistakes. Video is such a powerful medium, and it's getting more popular, yet more accessible, every day. Everyone can embrace video and become storytellers; the variety of available stories is limitless.

Older computers weren't good for heavy-duty, high-definition editing. You needed to use $1,200 programs such as Adobe Premiere or Final Cut Pro, but they were slow and crashed a lot. Cameras were often incompatible with certain programs, and they required copious cables and annoying tapes. It was an expensive mess.

Around 2008, you could film on cameras with disks, in universal format. That was a big shift. Still, it's nothing like today, where I can shoot with my 9.7-inch iPad Pro, in 4K video quality, with a 256 GB hard drive, and a processor that blows away my early computer—on this little slab of glass.

I can film an event in 4K (Ultra HD), capture great audio, do a quick edit in an app like Adobe Premiere Clip, *on my iPad*, then import it into Adobe Premiere Pro on my computer. I can then edit more intensely, use After Effects or Photoshop, and add green-screen work. In the end, I've created a professional-quality film in a fraction of the time it used to take. The workflow is cheap and simple. It's amazing.

Most people can edit fully on an iPad, which means they can share their videos with the world immediately. You can shoot in 4K, edit in iMovie or Premiere Clip, add titles and photos, and with a click of a button, it's on Facebook, YouTube, or wherever you want. You can upload a finished video while you're still physically at the set. You don't need to go home or to the office.

The BBC was one of iOgrapher's first adopters, for which I'm eternally grateful. They realized it wasn't necessary to spend fifty to sixty *thousand* dollars on a Red Camera, which is what most feature films are shot on. iOS devices get the job done almost as well. When using the Red Camera, files are massive—often 6K (6,000 pixels). Using a Red Camera is not even worth discussing unless you're producing a blockbuster film.

My students taught me an important lesson: what matters most is having a great story. They showed me the future; they don't even watch TV anymore.

They're on Snapchat, Instagram, Facebook, and YouTube. They don't care if the video is shot in 4K; they want a good story. They want to be entertained. iOS devices level the playing field and let the best stories rise to the top. Having the best equipment or the most money has become irrelevant. A phone is good enough. The film world used to be an oligarchy. Now, it's becoming a meritocracy.

At VidCon, the YouTube generation's biggest video convention, Howie Mandel approached my table. We had a nice, quiet chat, and nobody around recognized him. A few minutes later, a YouTuber named Justine walked by. You could swear she was The Beatles; people were screaming and chasing after her. She had bodyguards. Seeing a YouTuber achieve that level of fame convinced me that success is about quality of entertainment, above all.

Sure, I'd still use the Red Camera for a $100 million feature film, because it allows me to use a variety of lenses and do some digital intermediary things. For day-to-day storytelling, though, the devices in our pockets are more than we need.

The beauty is, mobile devices improve every year. If you can dream something, you can create it on video. The movie *Tangerine* was shot entirely on an iPhone 5, and it won the Sundance Film Festival, got picked up by a major distributor, and netted a worldwide distribution deal. It had a nice run at local theatres and is now available on iTunes. It was produced for almost nothing. The creators didn't even need permits. They had a story they believed in, and they went out, shot it, and brought it to life with an outdated iPhone.

Who's going to be the next Spielberg, Lucas, or Coppola? A kid with an iPhone can get started, today.

Filmmaking is only one medium; the next wave is live video. One of my favorite examples is an NFL team we've worked with to connect players with fans—chatting live on video before and after games. Back in the day, I would've loved to talk to Terry Bradshaw after the Super Bowl. That used to be a ridiculous idea, but now it's reality.

Live video allows you or your brand to connect with the 1.3 billion people on Facebook. You can do questions and answers, show them a new product, conduct a webinar, give them behind-the-scenes access, or anything that comes to mind. As with all video, you're only limited by your imagination, as long as you have that trusty, powerful device in your pocket.

One of the best ways to use live video is to stream live events. Perhaps it's graduation season, but Grandma lives in the Midwest and can't make it to California to see her granddaughter graduate. Mom and Dad could stream graduation to Grandma, via Wi-Fi, free of cost. They could also film their kids' football or basketball games, plays, or any event in school or outside. Live streaming is a great way to archive footage and to get the community involved.

Live streams can be professional in quality, too. By using apps like Switcher Studio (discussed in chapter 11), you can pair four iOS devices and do a live, multi-camera shoot. That means your son's football game can have the same variety of angles as *Monday Night Football* on ESPN, also broadcast live. Using multiple angles can be useful in a variety of settings, from football games, to journalism, to music videos.

My friend Michelle Dozois, a fitness expert, used to pay $10,000 a day to hire a crew to film her workout videos. Now, she does everything herself with four iOS devices. She can edit on set with Final Cut X or Adobe Premiere Pro, and by simply selecting the four angles, right-clicking, and hitting "synchronize," all of her footage is lined up professionally. It's automatic. From there, she can do a simple, multi-camera edit. It's quicker, cheaper, and can all be done by one nonexpert.

Filmmaking has been democratized. Anyone can create quality content simply and cheaply.

You can also use your iOS devices to become a mobile journalist, or iJournalist as I call it. Get out your iPhone, and plug in a couple of audio devices or a handheld microphone. You can create your own station, like I've done with iOgrapher TV. I go to conventions, walk around and interview people, and the end result looks and sounds as if it's on television. I can then upload my story to CNN iReports, if I used iMovie. If my story is good enough, CNN will pick it up.

The results of mobile journalism are professional quality. Recently, a BBC reporter interviewed CEOs as he traveled around Europe, bringing nothing but an iPhone and an iOgrapher with a few attachments. He didn't have to deal with border checks, where his standard camera would have been taken apart. It also made filming less intrusive, because people are more comfortable when they aren't getting a huge camera shoved in their face. It's more natural and inviting.

Mobile filmmaking is great for sports, too; teams like the Boston Celtics are using our devices for coaching. By using apps that we'll discuss in chapter 12, they can do things like instant replay and slow motion. A great app called Coach's Eye allows you to film individual athletes—say a golfer's swing—and draw corrective arrows on it to show the athlete later, in slow motion. Coaches can also add voice annotations and then email clips to athletes for study at home.

High-school football teams often use the app Hudl. Someone films at the top of the stadium and sends each play immediately via Wi-Fi to the Hudl server in the cloud. The coach can then open the clip on the sidelines, seconds later. It's the ultimate coaching tool, and it can be used at any level.

The options for mobile filmmaking are endless, but you can start small. All you need is a mobile device and the desire to tell a story. From there, you can expand as far as you'd like, to Hollywood and beyond. iMovie is already installed on your iPhone and iPad; just get out and start shooting and editing. That's all you need at first. Next, you can buy the affordable iOgrapher case, which will give you stability and lighten the load of holding your device with one hand.

I'm biased, but the iOgrapher case really is the ultimate introductory accessory for your iOS device, and it can be upgraded to turn your device into a portable Hollywood studio. The two handles allow you to hold your device easily and comfortably, getting steady video. The "shoes," which I'll explain later, allow you to add accessories. You can add lenses and film in wide angle, like in *Lawrence of*

Arabia or *The Hateful Eight*. You can add microphones for quality audio. You can put your iOgrapher on a tripod for balance. You can do all types of panning. You can add a shoulder mount for comfort and maneuverability.

If you're a beginner, start slow. You can build your way up easily and inexpensively. The cheapest microphones start at fifty dollars, and go up to the $300–$400 range. Anyone can get started, but the sky is the limit—an accessible limit.

Olivia Wilde, the famous actress, recently shot a music video entirely on an iPhone. She augmented it with lenses and other accessories, and it looked as good as any other music video. iPhones can shoot closer than big, expensive cameras, so she got angles that were previously unattainable. GoPros share that advantage, but they have their limitations, which we'll discuss in chapter 3.

Every day a new option or story is unearthed. One morning I woke up and saw a photo in a magazine of Steven Spielberg talking to Tom Hanks, holding the iOgrapher Mini. I assumed he was using it as a viewer, connected to his gigantic, $100,000 camera, but I was still amazed to see it in his workflow. iOgrapher is useful for everyone, from beginners to legends.

In this book, I'll show you how to make any type of video

your creativity desires, no matter who you are. You'll need nothing more than the iPad in your backpack or the iPhone in your pocket, and perhaps a few portable, affordable accessories. The world is your film set; go tell stories.

PART ONE

DEVICES

CHAPTER 1

IPADS

||||||||||||||||||||||||||||||||||

When using iOS devices for filmmaking, there's an option for every budget. For beginners, I recommend visiting the refurbished section on Apple's website. It's a shame more people don't use it; I buy all of my computers and iPads that way, unless I need to test a newly released device. You'll get older versions of great products, and you'll save 20–30 percent (or more) on prices.

If you're a beginner looking to stream video—perhaps your kid's Friday-night football games—you can start with the iPad Mini 2. The 16 GB hard drive is a disadvantage for some, but it's plenty for streaming, which only requires enough space for the app to run. You don't need to store big files. The iPad Mini 2 costs $229 refurbished, and by adding an iOgrapher case ($60), a tripod ($20+), and some Wi-Fi, you can produce a professional-quality video stream for around $300.

If you want to create short movies, photography, or journalism pieces, you can upgrade to the 32 GB version of the iPad Mini 2, for $269. The 32 GB hard drive has enough space to store small files.

It's tough to beat Apple's refurbished section, but you can also check retailers like Target. They'll sometimes give you a $100 credit, or something similar, for buying an older device. You can use the credit for a tripod or other accessories. The iPad Mini 2 is old, but it's a fantastic device, as are all following editions. I still use the iPad Mini 2, because I have tons of them, and they work perfectly. If you're on a budget, a refurbished iPad Mini 2 is the perfect place to start.

You can even get by with original iPads. Their cameras are solid, and there are great apps (like FiLMiC Pro, explained in chapter 10) that can enhance video quality. A lot of schools still use the original, full-sized iPads, and they love using our iOgrapher products to upgrade their quality. Any iPad, except the first version that didn't have a camera, can be used to make quality content.

The Boston Celtics' strength-training coach uses an old iPad to film player workouts. He doesn't need 4K resolution; he just needs clear picture.

Basically, you can reinvent your old device and keep your old iPad forever. Add a couple of lenses, a tripod, and a microphone, and you can do anything. You can connect with your business's customers through live video. Walk around your show floor or factory. Take your old iPad, add an iOgrapher and some accessories, and stream live to your Facebook page that has all of your clients.

I tell my students, and everyone that I meet: "We're all documentarians. We're documenting our lives every day, whether we're using Snapchat or Instagram, texting a photo to Mom, or filming our kids playing in the backyard."

A few months ago, a Star Wars fan got a Chewbacca mask as a gift, and she filmed her own reaction. Three hundred million views later, she has paid off her children's college tuition. She scored a free trip to meet J. J. Abrams. This golden age of content is like panning for actual gold in the Old West: there's massive opportunity, but you don't know where the gold is. You don't know which videos will go viral, so you keep documenting and hoping for the best. All Chewbacca Mom needed was her iPhone and YouTube.

Every day, my son watches a YouTuber who makes videos where he talks about Minecraft. He has eleven million subscribers and is making millions of dollars a year. The opportunities are mind-boggling. Are you a storyteller? A musician? One of my friend's daughters has an incredible voice, so I've insisted she post regularly on YouTube to get her name out there. She goes to the University of Colorado and makes great videos of original songs and covers. We set her up with a bunch of iOgrapher gear, and now she's reaching an audience and getting people to notice her. Who knows where that will head?

As I've said, you don't need much to get started. Buy a refurbished iPad Mini 2—the $229, 16 GB one for live streaming, or the $269, 32 GB one if you want to shoot some regular video. Then, make sharing stories a habit. You'll improve every time you hit "record."

Because you can stream to certain channels that capture your files, you don't need to store much on your devices—16 GB can be plenty. If you're streaming out to YouTube and turn on the WireCast Go app, all of your finished content will be stored on YouTube for you to download to your computer later. That can be done by using free downloader software, like YouTube to MP4.

When you're done, upload your videos to Facebook, Instagram, YouTube, or anywhere else you'd like. The beauty of social media is that your content can be repurposed endless times. For $229, you're in the game, and you can do serious filmmaking.

For people with slightly bigger budgets, I recommend the iPad Air 1 and 2. They're great devices with bigger screens and better processors. The original iPad Air (16 GB) costs $279. You can use it for everything I've mentioned, with the added benefit of a bigger screen for filming, or you can use it as a teleprompter. The main advantage of the iPad Air is its bigger screen; it's mainly an individual preference. I recommend the iPad Mini 2 to most beginners, but both devices have 1080p, HD cameras, and both create quality content. If you want the bigger screen, the iPad Airs are great.

By adding the $10 FiLMiC Pro app, explained in chapter 10, these sub-$300 devices will shoot in higher quality than you'll ever need.

For 99 percent of people and uses, you'll be set for life with one of those two devices. With the 32 GB versions, you'll have enough space to tell stories, run some apps, and do anything. Buy one of these reasonably priced devices, and you have all you need to start making content.

The only real negative is storage space. A hard drive of 32 GB will get you about two hundred minutes of video, maximum. The exact number depends on the settings you're using, but for fifteen megabits per second, that will be the case. Regardless, your hard drive is small, so your filming time is somewhat restricted. Technology tends to fix most problems, though, and companies like SanDisk are beginning to make wireless hard drives. If space is an issue, you can use a wireless hard drive, or a plug-and-play one.

All of Apple's refurbished devices come with a one-year warranty, and if you want, you can get the extended Apple Care warranty for a few dollars more. I've never had a problem with refurbished devices, but I'm meticulous with my stuff. Apple wants you to have the best user experience ever, and they do a fantastic job, but always be careful.

Right now, I'm looking into my new, 27-inch iMac 5K retina display. I saved $1,200 buying it refurbished, and it works great. You don't always need the latest and greatest, and you'll still have more capability than you need. A person with a refurbished iPad can make better content than someone with a $100,000 camera. All that matters is the quality of storytelling.

Some people *do* want the latest and greatest—especially

power users like me—and that's fine. If so, get the 9.7-inch iPad Pro. It shoots video in 4K, which is huge. The 4K was introduced on the iPhone, but it has extended to the 9.7-inch iPad Pro.

Another option would be the new iPad Mini 4, which is larger than the previous version. It has a nice retina display and a thrilling 128 GB hard drive. For $599, you could create anything your heart desires.

I'm crazy, so I have the 12-inch iPad Pro, which can be a laptop replacement. I love it and bought it on release, but it's gigantic.

The 9.7-inch iPad Pro killed the iPad Air, because it has the Pro guts in Air size—a 4K camera and 256 GB hard drive. The 4K video files are gigantic, so you'll need that extra storage. Facebook, Instagram, and the like don't currently support 4K, but filming in 4K allows me to edit, pan, scan, resize, and so on without losing resolution. I'm mastering for the future. Because I shot in three times larger than 1080P, I have leeway. The 9.7-inch iPad Pro is the best tool you can buy, but the Wi-Fi version costs about $900.

$900 isn't cheap, but the 12-inch iPad Pro can replace a computer, broadcast camera, and DSLR. If you're a serious filmmaker, it's a steal. You can do *everything* on the $900 iPad Pro, using professional mastering tools like you're in Hollywood.

For *really* serious filmmakers, you could get four iPad Pros and do multi-camera shoots. They wouldn't have to be the 256 GB ones, but let's say they are. Your four 9.7-inch iPad Pros would cost $3,600, and then you could add a 12-inch iPad Pro ($700) to use as your switcher. If you then bought a few iOgrapher cases, tripods, and microphones, you could have a four-camera, Hollywood-quality workflow that shoots in 4K, for a ridiculously cheap price.

I recently used that setup for my fitness-trainer friend, Michelle. I set up the four smaller iPads as cameras, then the bigger one as my switcher, using the app Switcher Studio (explained in chapter 11). I set each device to my desired quality, 4K in most cases. Then I used the switcher iPad to master the other four devices and to switch back and forth between angles. When I was done filming, I had a finished video I could share with the world, on the spot. I was also able to stream live while shooting.

To keep everything in 4K, all I had to do was import the video later into an app like Adobe Premiere Pro or Final Cut X. I could then do a multi-camera edit, mastering in 4K, and then upload it to YouTube (because YouTube

> ### What's a plate?
>
> A *plate shot* is used to facilitate special effects. You film without any actors or specific set pieces—it's a pure background shot. You shoot with the same lighting as the "real shoot," and then use the plate shot as a reference to ensure that there are no discrepancies in the video after adding effects. A plate shot ensures consistency—so that your film looks realistic and consistent.

allows 4K files). The ability to do all of that for under $5,000 is *insane*. You used to need multiple cameras that cost $3,000–$5,000 *each*, plus cabling, plus a switcher. We're lowering the price bar, yet upping the game.

The 4K is a game changer, because when I shoot in 4K, it maximizes my options for postproduction. If I've got a great, wider-angle shot that I want to zoom in on, I can do so and edit without losing resolution. Then I can upload it on Facebook, YouTube, Instagram, or anywhere else.

Another example would be if I was shooting a plate for a visual-effects shot I want to include in my movie. By shooting in 4K with my iPad, I'll have more pixels to play with in creating that shot, which will help in postproduction. The 4K is a great way to master your content. The only negative is that the files are enormous, which means you'll need to be careful. Fortunately, my computer has a 3 TB hard drive. That cracks me up, because I used to think my old, 125 GB hard drive was massive. Heck, I remember *megabyte* hard drives. Where will we go next? Nobody knows.

You may need to down-res your 4K video for certain outlets, but that's easy. If you're editing on your iOS device, iMovie or Adobe Premier Clip will automatically down-res to 1080p, the high definition you're used to from TV and Netflix. You won't see the power of 4K until you bring the file into an editing program like Adobe Premiere Pro or Final Cut X, on your computer, where you can really resize things, play around, and see how things remain crisp and clear.

If you have the 256 GB iPad Pro, you'll have plenty of room to shoot in 4K. I wouldn't shoot a two- or three-hour football game in 4K unless your boosters needed a Blu-ray DVD, but you should use it for your own content—You-Tube shows, and so on—things that are five, ten, fifteen minutes long.

As you can see, the sky is the limit. On the budget side, though, there are more options. At iOgrapher, we love hearing stories of people reinventing old iPads that were used only for playing Angry Birds. People don't realize the untapped power old devices have. By adding an iOgrapher case, and perhaps a wide-angle or telephoto lens and a microphone, old devices turn into powerhouses. We recently sold some stuff to Boeing, where they're using old iPads to shoot videos of planes they're fixing, to share with other mechanics later on.

I was introduced to mobile filmmaking with the old, full-sized iPad, and I was amazed. It was obvious it needed a case like iOgrapher to reach its fullest potential, but I immediately saw the potential to capture moments on the fly. As I've said, iOS devices are so intuitive. Anyone can use one effectively. DSLRs, on the other hand, are like opening Pandora's box. You need to know aperture, exposure, ISO settings—if you aren't a photography expert, you're lost, and you can't use the power of the DSLR.

I've played with DSLRs for years, and I still have problems. Occasionally we need super-high-resolution for print, so we use a DSLR. Recently, I thought my DSLR was broken. My shots had dark edges, and everything looked weird. I was digging into aperture settings and playing around, and it turned out that someone had put the camera on "toy mode." It was a preset that someone had turned on accidentally. Even veterans can get lost.

Not with the iPad—you open it up, and it focuses on your focal points automatically.

There is a caveat—you need to lock your focus. Otherwise, you'll get a pulsing action that will ruin your shots, because iOS devices constantly scan for focal points, and those can change. If someone walks behind you, the camera will say, *Let me focus on that.* To learn how to do this, watch our tutorial at bit.ly/lockfocus.

iOS devices are perfect for a classroom. When I was teaching, I'd set up a green screen and the iPad, and the kids would go up, shoot, and AirDrop the files from the iPad to their computer. They were able to edit seconds after filming. They didn't need any cables, and it was a huge time-saver.

The only negative was that we saved so much time that I needed to come up with more lessons! It was more work for me, but more reward for students.

All jobs have *some* role for video, so I believe the ability to create videos with such little friction is valuable for everyone. If you're in marketing, management, or education, you've got to train your employees, which means you should know how to create videos for Instagram, YouTube, Twitter—whichever platform is most useful for you. People will always need the ability to make video tutorials and presentations, and iOS devices give you the lowest entry point of any tool. They're amazing; anyone can use one.

When my son, Alexander, was four years old, he loved LEGOs. One day I set up the iOgrapher and iPad on a mini tripod and opened a great app called iStopMotion. It's a simple app that allows you to take frame-by-frame animation and automatically turn it into a stop-motion animation (explained in chapter 12).

I showed him how to use the app. Click here and, boom, it takes a picture. Now, let's move this LEGO a bit this way. Keep doing it. I played ten frames or so back to him, and he saw this magic happen. It's an awesome way to teach motion and time. You don't want to go from A to B in a giant step; you want tiny pieces in each frame so that you have this nice, smooth motion.

He *got* it. He was an alpha male, even as a little monster. He pushed me out of the way and was like, *I get it.* He clicked and clicked and made these fun little stories, showed his friends, and they started doing it. It was contagious. They're currently using it at the local kindergarten. I'm thrilled; it's a great way to get kids to have fun while learning animation on an iPad. Kindergarteners are learning things adults previously didn't have access to, and it's an amazing time to be alive.

CHAPTER 2

IPHONES

〰〰〰〰〰〰〰〰

We made our first iOgrapher for the iPad, and we initially thought it would be our only case—did people want to use a small iPhone screen for filming? Then the BBC approached us, asking for an iPhone 5 case, and we had a lightbulb moment. They had loved the iPad case, but they wanted the portability of an iPhone to use for their World Cup coverage in Brazil. They took our cases to Brazil, shot some great videos, and couldn't stop raving about how close to the action they were able to get. The floodgates had opened; iPhones weren't too small for serious filming.

In fact, the BBC proved that the iPhone's lack of size is actually an advantage. They also loved that they could quickly edit on the phone itself, allowing them to get content out immediately to viewers. In the new world of instant information, using an iPhone is a competitive advantage over other news outlets.

The iPhone 5 was the first iPhone to have a great camera— you could shoot real, professional content. Bentley shot a commercial with an iPhone 5, getting shots they never could have gotten with a normal, immobile camera; it really displays the inside of their cars. It was a beautiful commercial and used no additional cameras.

The Golf Channel hopped on the iPhone bandwagon, too. They felt it was the first mobile device that could be used in a professional setting. As I mentioned earlier, Sundance Film Festival winner *Tangerine* was shot entirely on an iPhone 5s. The point is, we want to stop people from using their phone just for selfies. Your phone is so powerful—you can stream through Periscope, Facebook Live, or YouTube Live; you can create quality, broadcast journalism (as an iJournalist); or you can do anything else your creativity suggests.

I used to shoot almost everything on my iPhone 6+. It has a 128 GB hard drive. It shoots in 4K. If I want to jump on a Facebook Live or Periscope broadcast, I put on my iOgrapher case, plug in my microphone, put it on my tripod, and I'm off to the races. This past weekend, I was at a convention. I popped in my handheld, wireless mic and went out and interviewed people. The versatility, mobility, and quality are remarkable.

The biggest game changers were the iPhone 6s and 6s Plus, which have given you 4K video capability in your pocket. The only thing stopping you, literally, is your creativity. That goes for content in any genre—sports, coaching, journalism, filmmaking, animation, photography—you name it.

Apple recently stepped up its game and came out with the iPhone 7 and 7 Plus—now I use my 7 Plus for everything. It has a 256 GB hard drive, and the awesome addition of a digital zoom and *dual-lens capability*—allowing for wide-angle capability. By adding an app like FiLMiC Pro (described in chapter 10), you can do cool things like simulate a depth-of-field look. The new iPhones are remarkable.

They did remove the headphone jack—but we'll address that minor issue in the following chapter.

Young people—heck, everyone—want to be creative. Do you have an opinion about the current elections, and think you can make a change? Start a Facebook page, broadcast on Facebook Live, take comments—now you have your own news show. There's no way to list all of the possibilities, because they're endless, but each individual can play his or her unique part.

There's a reason that at every news conference on TV, like in election campaigns, you'll see a sea of people holding up iPhones, filming. There's no need to bring a big camera; everyone already has a quality camera in his or her pocket. If people brought expensive DSLRs, they'd need to go back to their computer to edit, and only then could they share their content with the world. By then, it's old news. In the time it took you to get home, I sat on a bench and did a quick edit on my iPhone, added some titles, and, boom, the world saw it on Facebook, Twitter, Instagram, YouTube, and CNN. You're still working, but I'm on to the next story.

The fact that you have *that* much power inside your pocket, with an iPhone, is amazing. I apologize to Android users—I've always had iOS devices, so I can't comment on Androids. I like that my iOS computers interact with my iPhones and iPads—I prefer the iOS workflow to Android. It's personal preference. I have nothing against Android users, and you have the same potential for creativity in *your* pockets.

When I founded iOgrapher, I had no idea how I was going

to market this *thing*. I just knew that schools would like it. Then I randomly heard about a broadcast journalism expo in California. I went around with a little monopod and my iPad Mini on top, and people swarmed. *Univision, Telemundo,* CNN, and all these local stations needed to know what this *thing* was.

Once we started using iPhones, journalists went nuts. Now when they're filming at a fire, or somewhere they can't get their satellite to post footage, they can send their video file over their iPhone's LTE connection. It goes into the cloud, where people back at the station can get it out, live, right away. Journalism is about speed, and filming from an iPhone is the ultimate advantage.

The nightly news is dying. We can get information on Twitter and Snapchat so much faster. How much longer will the nightly news even exist?

I've talked more about iPads, but if all you have is an iPhone, you can do everything I've mentioned, and more. You can call your girlfriend or mom, schedule meetings, return emails, and then, oh, by the way, flip on airplane mode, and shoot your feature film that you'll submit to Sundance.

It's mind-boggling. There's so much power in iPhones;

they're like having a high-end computer, professional camera, editing device, and distribution network in your pocket. An iPhone is literally all you need to be a successful, award-winning filmmaker.

CHAPTER 3

OTHER DEVICES

|||||||||||||||||||||||||||||||

The iPad and iPhone are all you *need*, but there are of course other devices that are useful for certain people and scenarios. You've probably heard of the GoPro camera. When it came out, I loved how tiny it was and the fact that you could put it anywhere (like you can now with an iPhone). I disliked, however, that the camera's superwide lens caused a crazy jerk in your footage if you turned your hand at anything quicker than a snail's pace.

The iOgrapher has "shoes" (screws to insert accessories) on top, to which you can attach a GoPro. The best way to use a GoPro is to put your iPhone in an iOgrapher case, then attach the GoPro on top. You can then use the iPhone as a monitor, to keep an eye on potential jerkiness and to set up your shots. Until recently, GoPro models didn't have their own monitor, so otherwise you're forced to guess, which often leads to disaster. You don't know that your shot has been ruined until it's too late. Using your iPhone as a viewer unlocks the GoPro's power.

You can also open the FiLMiC Pro app (which I'll detail in chapter 10), and possibly attach your telephoto lens. Then you can get a close-up of the same shot (a wide shot of the top shot), and now you've got two different shots to bring in for edits later. It's a great way to combine forces. I love the GoPro when it's optimized.

The GoPro Hero 5 came out recently, and it is a massive improvement. It comes with its own little LED monitor on the back, and it also has voice activation. You can set it up and then say, "Start filming," and it will. You can also set it to record up to 240 frames per second, for slow-motion shots. It's still useful to attach your Hero 5 atop the iPhone via an iOgrapher—use the iPhone as your close-up shot, and the Hero 5 as your wide-angle shot. They complement each other perfectly.

The BBC does something similar, but with an expensive ($10,000) Canon C300 camera. They put the iOgrapher iPhone case on top of the camera, on the shoe, and take a wide-angle shot with the iOgrapher, while getting a super-close-up shot with the Canon and some fancy lenses. Everything complements each other perfectly. There are tons of options for different scenarios.

GoPros are action cameras created by a former surfer who wanted to get surfing shots, and they've turned into a phenomenon. Their plastic housing makes the camera waterproof, and you can put them anywhere. The new one, the Hero 5, doesn't even need plastic housing—it's waterproof without it. All of the new ones shoot in 4K format, too, so you can shoot some really great, super-slow-motion footage with them, and have some fun. The only problem is they're hard to maneuver. Any little movement, because the lens is so wide, and there's a giant, unwanted on-screen movement. Also, you can't see what you're shooting, unless you add an iPhone with an iOgrapher to use as a monitor. Otherwise, you point and hope. Hence the solution mentioned above.

GoPros also have an adapter, to which you can add a microphone. By adding an iPhone or iPad with an iOgrapher case as a monitor, then putting a shotgun mic (detailed later) on top of the case, you'll get great video *and* audio. All of a sudden the flawed GoPro will create amazing film. When you're done, you can wirelessly transfer the footage to your iOS device for editing, through the GoPro app.

The GoPro is a great backup camera when used properly. By using an iPhone as your monitor, you can start and stop, change formats, and do anything you'd want, via the GoPro app on your iPhone. And with the new Hero 5's voice-activation feature, shooting two different angles—one with an iPhone, and one with the GoPro—can be done seamlessly, without the need to fiddle.

We also have a case for the GoPro itself, called the iOgrapher Go, that is compatible with other action cameras like the Sony Action Cam, the Canon VIXIA Mini, and the Black Magic Micro Cinema Camera, a professional-level camera.

I personally like shooting with my GoPro, but only when I have the same stability I'm used to with iOS devices and iOgraphers, plus the ability to view what I'm shooting. By using our iOgrapher Go case, and an iPhone or iPad with an iOgrapher, I can accomplish that. I also will add a face plate, made by the awesome company Back-Bone.ca. That allows me to add DSLR or PL-mount lenses to my GoPro, so I can get insane footage. The GoPro is a great but flawed base camera. When you optimize it with accessories, it's legendary. I commonly use my GoPro as a backup camera to shoot an extra wide shot as an extra option for postproduction. *(Note: At the time of this writing, BackBone doesn't yet support the GoPro Hero 5.)*

The GoPro is also great for live streaming, especially on Periscope. You talk into your iPhone, as usual, but place your GoPro somewhere else, as an extra angle. It could be ten feet away from you, and you can switch to that angle via the Periscope app on your iPhone. By using the GoPro, you get a wider-angle shot, which prevents the viewer from feeling as if he or she is staring you in the face, which is how it looks when you live stream with the standard, narrow iPhone lens.

Your GoPro connects to your iPhone through Wi-Fi, like most external devices or cameras with apps—so when you're done shooting, you open the app, review the footage, press a button, and it transfers automatically. It's easy and doesn't require any cables, computers, or card readers. You can do everything on the spot.

There are a lot of GoPro users out there, and my goal is to make their experience better. The GoPro is a great camera, but it only reaches its fullest potential when you add stability, upgrade the audio, and monitor what you're shooting. Once you do that, you can create great content.

Next, we'll discuss the Osmo, created by DJI, one of the world's top drone companies. For less than $700, the Osmo shoots 4K video, and it can be used as a gimbal. A gimbal is an incredibly valuable device that takes fluid footage while you're moving. For example, say you're walking straight ahead, and I'm following you, filming. If I want to come around you and get in front, I *could* do that with my iPhone or iPad, but it won't be nearly as smooth as when shot with a gimbal. The gimbal allows me to lock you in as a person, so when I walk around you, the camera will smoothly follow you in a circle, capturing fluid, beautiful footage.

My customers asked me to create my own gimbal, but luckily I found the Osmo instead. It's perfect, and all I had to do was make an Osmo attachment for our iOgrapher cases. A gimbal is a stabilization device. iPhones have their own stabilization, and it's solid when you add an iOgrapher case, but the result isn't nearly as beautiful as when shot with the Osmo. The Osmo has its own handle, so you'll have three handles when you add our case. It doesn't have a place to put a microphone, but our case solves that problem, so you can get great audio, too.

The Osmo allows you to get the same quality footage as a professional cameraman with thousands of dollars' worth of gear, for a fraction of the price. The footage is in 4K, which allows you to master it and do amazing things. By using the Osmo, your film will be far more cinematic. It's a great addition if you're doing wedding videos or short films, or any time you want to do professional-quality camera movement for an amateur price.

It's a new tool, so I don't yet know many people who use it. That will change soon. If you're a serious filmmaker, you should add the Osmo into your workflow. I bought it solely as an experiment, but that's been a theme of my life—whims and accidents turning into great things. It's a great mind-set to have.

When I played around and paired the Osmo with the iOgrapher, microphones, and lighting, the ability to make things move fluidly was a dream come true. Adding accessories makes an already great product improve a hundredfold. Always optimize the potential of these powerful devices. I take my Osmo with me everywhere, now. It's part of my workflow, because when I want that shot, I've got it handy. Try it out, and I'm sure you'll be impressed.

PART TWO

||||||||||||||||||||||||||||||

AUDIO

CHAPTER 4

MICROPHONES AND CONVERTERS

The old adage says, *You can have bad video, but you can't have bad audio.* People will watch a flawed video if there's a good story...as long as they hear everything clearly.

Mediocre video might bother viewers, but if they understand and like the story, they'll continue watching. If they can't hear the audio, though, they won't understand the story and will give up. When video is poor, you can close your eyes and still get the point. Audio has many potential problems—too low, too garbled, or a million other things—and those problems render the viewer clueless.

Failed video can be turned into a podcast, but video without audio is useless. Producing professional, high-quality audio is essential, and the way to do that is by adding a microphone to your workflow.

When I was teaching, a lot of "helicopter parents" would help their kids with projects. One mom was insistent that her son look and sound professional for our news broadcast team. I realized I had to figure out how to add microphones to iOS devices to optimize audio. Fortunately, the big microphone companies like Rode and Sennheiser have started to make things specifically for iOS and have upped the game significantly. They've made it easy.

While the onboard iOS microphones are passable, I recommend everyone add microphones. We've searched the stratosphere to find the best and least expensive ones to sell on our website. We've also found the most expensive and professional ones. You can start with a basic microphone that costs fifty-nine dollars, and it'll sound ten-times better than your iOS device. That said, each version of iPhone and iPad gets better microphones, so this could change. Eventually, an iOS microphone may be sufficient, but not as of this writing.

The biggest problem with current iOS microphones is that they're built in, so they pick up everything in front and behind you—noises of all kinds. That's good enough if you have no other option, but spending a few bucks for entry-level shotgun or lavalier microphones will be a considerable improvement.

Microphones come in a variety of types, but we'll start with the long, directional microphone called the *shotgun mic*. You've seen them on TV—wherever it points, the long tube will pick up audio. It tends not to pick up audio to the right and left of it, called surrounding audio. Shotgun mics are directional mics, great all-around microphones for shooting concerts and interviews, and for other times when you need audio from one stationary spot. Everyone should have one in their toolbox.

Next is the *lavalier mic*. They're used in many different situations, from home marketing videos to big movies. They're tiny black pins you can see, usually attached to a long wire, which plugs into your device. We use one called the iRig Mic Lab. It plugs into your headphones for monitoring while filming, so you can tell a person to speak louder or softer. Lavalier mics are great for close-up situations, perhaps when you're interviewing someone and only want his or her audio.

Handheld mics are what you would see on the 11:00 news on TV, if you still watch it. The anchors are holding it, and it says "ABC Channel 7," or "ESPN." They're typically used for broadcasting and interviewing. You can use them for karaoke. They give you great sound and are perfect for iJournalists going out and telling stories in the world.

Boom mics are often used in movies. You have a *boom pole*, which is like long fishing pole, to which you attach a microphone and connect the cable to your iOS device. You use a boom mic when you don't want people to see a microphone. You want great audio, but don't want people to see a lavalier mic sticking out. Or perhaps you're doing action filming, but a lavalier would rub up against the person's clothing and ruin the audio. You can hold the boom on top of the person you're filming, or point it at them, and it'll pick up his or her audio. Boom mics are a timeless tradition in Hollywood. They are how audio gets into movies and TV, and now you can do the same from your iPhone or iPad.

Last, but not least, are *wireless microphones*, which vary greatly in price and quality. One great company, Samson Tech, has started to create sub-$100 solutions, lavalier or handheld wireless mics that plug into your iOS device and let you walk around, unencumbered. They sound great; I recently used one to interview people at a convention. You may get some feedback if you're near someone on the same signal as you—you can pick up interference. Your

LTE signal on your iPhone could interfere, too. It's a great option if you're aware of the minor, potential drawbacks. Another way to limit feedback is to put your device in airplane mode, so that it isn't signaling the LTE connection, which sometimes gives static.

Beginners should start by purchasing a basic shotgun mic. On our website, we sell the Rode Video Micro, which costs fifty-nine dollars but picks up great audio. It comes with a dead cat—a fuzzy microphone cover that stops wind from affecting your shot. It fits into your backpack or purse, picks up great audio, and the price is mind-boggling. The Rode Video Micro is the perfect entryway microphone.

Next, purchase a solid lavalier mic, like IK Multimedia's iRig Mic Lav, for $45. It fits in your pocket and gives great audio—perfect for a live stream on Facebook or Periscope. You can add a headset to monitor audio, or plug other iRig Mic Labs into it, creating what's called a *daisy chain*. Basically, you can plug your main microphone into your iPhone, then plug another one into that microphone, through the headset jack, and now you've got two microphones. You can keep doing that until you've got twenty microphones, or as many as you'd like.

Those are the two perfect starter microphones for anyone. They're all you need for entry-level filmmaking.

From there, you can graduate to the $250 Rode Video Mic Pro shotgun microphone, which has more bells and whistles. It's more of a professional tool, with better quality, and the ability to change pre-amp volume to -10db for noisy environments and +20db boost to reduce camera noise.

For lavaliers, Sennheiser, an amazing company for professional-level stuff, has teamed up with a company called Apogee. They have a high-end, $175 lavalier mic that has a mixer inside and makes amazing audio.

Be warned, though: you get what you pay for. When I started the iOgrapher store, I decided I'd only sell brands that were the best and most tested. I've heard too many stories from people who bought a fifteen-dollar shotgun mic on Amazon, and cried to me later when it didn't work. Small companies send me samples all of the time, but they inevitably fail to compete with the big boys. Fortunately, big companies like Samson, Sennheiser, Rode, and IK Multimedia seem to get it. They know they need to make accessories for the masses, not just professional filmmakers. Everybody wants to make movies, and now everybody can.

Converters are an important piece to the puzzle, too. The iPad, iPhone, and other portable devices, like Androids, have a special connection called a TRRS. Because of that, anything you connect to them must use that same signal. Many microphones on the market were made for DSLRs and big broadcast cameras, so they have a TRRS or XLR connection. They're not compatible. Fortunately, all you need to do is add a converter.

Many current mics are being made TRRS-ready, for mobile devices, but if you have an older one, you'll need a converter. I usually use Rode for cabling. One is called the SC-7, a patch cable. It has a TRRS connection on one side, and TRR on the other side—your mobile device plugs into the TRRS side, and your microphone plugs into the TRR side. Problem fixed.

Rode also makes a great converter called the SC-6. It goes into your iOS device, but has two lines of audio going in, so you can plug in a headset to monitor your audio.

Many schools have old microphones; they're usually XLR. That was the old, professional standard for big broadcast cameras. IK Multimedia's iRig allows you to plug that big, XLR cable into one end, then the TRRS connection into your iOS device. This allows you to use an old, quality mic with your iPhone. The iRig 3 allows you to change levels, loudness, and so on, and they're upgrading it soon. It's a great way to reinvent your old microphone, especially if you're a school with a limited budget.

The Apogee Duet is a great product by a great company, on the higher end of the price spectrum. You can plug the Duet into your iOS device's lightning-jack plug, then plug 2 SLRs into the Duet, which you can then plug into a soundboard. It's a powerful device with many potential uses, and that's just one.

If you followed my advice and bought the Rode Video Micro mic for your iOS device, you'll need an SC-7 cable to make it compatible. The lavalier mic I recommended (iRig Mic Lab), however, is compatible as is. Therefore, you'll only need one cable for that shotgun mic.

Apple's decision to remove the headphone jack in favor of the lightning jack (what you use when you charge your phone) on the iPhone 7 and 7 Plus complicates things, but there is a simple fix. They also created the lightning-to-headphone-jack connector (available on their website), so that you can plug your microphone's 3.5-millimeter cable into the adapter, and then into your new iPhone 7 or 7+. That said, if you're planning on using the iPhone 7 or 7+, you may want to look into lightning-jack-ready products, like Rode's iXLR adapter. It allows you to plug other types of microphones, like shotgun or lavalier mics, into your new phone via the adapter for excellent audio.

I know this can be confusing, so email me at davesbook@iographer.com with any questions on mic compatibility, and visit the *iOgrapher* blog for the latest updates on new technology.

CHAPTER 5

MIXERS

In addition to a converter, you may need a *mixer*. By default, your iOS device captures audio through the headphone jack (or lightning jack, on the new iPhone 7 and 7+). If you plug in a microphone, as we've discussed, it will sound great, 90 percent of the time. However, audio coming from the headphone jack is a *mono track*, meaning only one line of audio can come in. When you need more than one line coming in, you're in trouble. To fix it, you need audio that comes through a *stereo track*, which Apple's headphone jack can't do. (Audio coming through the lightning jack is *stereo*, so you iPhone 7 and 7 Plus owners can ignore this.)

Shows like *Seinfeld* are a perfect example. Each character

was either wearing a wireless mic or had a boom pointed at him or her, meaning they all had their own, committed microphones. That's intentional, because when you use separate microphones, audio levels can be fixed in post-production. If George speaks too low, and Elaine too loud, that can be easily fixed. If everyone had spoken into the same line of audio, like what happens on your iOS device, that problem couldn't have been fixed. I imagine apps will soon be created to fix this problem, but for now, you need to buy a mixer. A mixer allows you to harmonize the audio levels of multiple incoming voices, and switches the audio from mono to stereo.

As I explained in chapter 4, beginners should start with a shotgun mic or lavalier. You'll get great audio, as long as you're doing something simple, like Facebook Live, where only you are talking. When you're ready to graduate, you'll need to address the problem of bringing multiple incoming sources of audio into your iOS device. The best way to do it is the same way they do it in Hollywood: by using a mixer. The prices are surprisingly reasonable.

Apple is starting to tinker with devices that can be plugged into the lightning jack (such as the new iPhone 7 and 7+), which does support stereo tracks, so the future is bright when it comes to sound. For now, though, non-iPhone 7 and 7 Plus owners will need a mixer.

My number-one choice for using mixed audio (switching mono tracks to stereo), is the MXL Mini Mixer, from Marshall Electronics, a longtime maker of professional microphones for musicians. They were the first company to create a mini mixer, which accepts four lines of audio, then plugs into your iOS device. (Note: The mixer doesn't do any conversions, so the microphones must be TRRS-ready. Otherwise, you'll need a converter, as explained in chapter 4).

Here's an example. I could plug my MXL Mini Mixer into my iOS device, which enables four potential inputs for microphones. I can plug my shotgun mic into one, and place it atop my iOgrapher case. Then I can plug in three lavalier mics, and attach those to my three interviewees sitting in front of me. Or maybe there are two people, and one can go on me, because I'm interviewing the person from behind the camera. This setup affords me great options, and I can have my headphones plugged into the mini mixer in order to monitor and fix the audio levels while filming. The mixer has individual gain controls, so if the audio is coming in too hot, I can lower it. I can completely control the volume levels of what I want the iPad to capture. It's not the best-looking thing in the world, but $229 is a bargain for a quality mixer. It's a small rectangle that could fit in your pocket, yet it connects to your iOS device, and accepts four microphone inputs, plus headphones. You can also plug it into your PC via a USB cable.

Next, we have Apogee, a big company that has been making great products for musicians, for years. They're famous for making high-quality mixers that musicians plug into their guitar on one end, and a microphone on the other. The sound goes to GarageBand or whichever editing tool they use, so the guitar sound and singing can harmonize. They've expanded to allow multiple microphone inputs, which is perfect for filmmaking. I bought the Apogee Duet when it came out a couple of years ago, and I've had a great experience. I once had a problem, sent my Duet in, and a couple of days later, I received a brand new one. It costs $600, but I've gotten what I've paid for.

The Apogee Duet is a step up from the MXL Mini Mixer. It allows you to connect XLR cables, which is perfect for connecting to a soundboard for an event. When I'd film football games, our announcers were using professional, ESPN-level headsets that had XLR cables attached. By using the Duet, we were able to run long XLR cables to the top of the stadium, in the broadcast booth, and get professional-quality audio. The Mini Mixer doesn't support XLR cables. Like the Mini Mixer, the Duet has individual gain and loudness controls, and you can monitor the audio. It's also great for educational use or school plays. It has tons of options, and if you can afford a more professional-quality mixer, the Apogee Duet is the perfect choice.

I know some people who are starting their own production companies, and in the past, they would've needed tens of thousands of dollars' worth of cameras and editing tools. Now, they can buy a few iPads, microphones, and a professional mixer like the Apogee Duet, and they're set, for a fraction of the price. This is the way of the future.

Apogee came out with an app recently, the MetaRecorder. I use it with my gigantic, 12-inch iPad Pro, because I refuse to make an iOgrapher case for something so big. When the MetaRecorder app is running, I can capture audio from other iPads, via Wi-Fi.

If I was working at a school with a limited budget, my students could all use their own iOS devices, and the MetaRecorder app would send all of their audio onto my master iPad, via Wi-Fi. The app turns your master device (my iPad in this example) into a mixer, while turning each phone into a bootleg (but quality) wireless microphone. If the students then added a microphone to their iPhone—say, a lavalier—I would get incredible audio sent back to my MetaRecorder mixer. Then, I could export the file to Final Cut Pro or another editing tool. The app is free, or five dollars for the advanced version.

The MetaRecorder app is perfect for media teachers, because they can spend their limited budgets on quality microphones, rather than buying a ton of iPads. Most kids already have iPhones, so that's unnecessary. Buying microphones instead is the smart, efficient thing to do. The app is intuitive, and when combined with quality microphones, it's the perfect way to teach students professional-quality audio.

Finally, there's the Zoom line of products. The one I use constantly is the Zoom H-6, which has the ability to capture up to six lines of audio. That's perfect if your school has old microphones that aren't compatible with iOS devices. Most schools have old, handheld mics that have XLR cables. By using the H-6, you can repurpose these old microphones, which do still give great audio. The H-6 allows you to plug in four different audio lines, or five, if you add an adapter.

When I first started teaching, all my school had was those old handheld mics. I wanted the video quality of iOS devices, but I needed better-quality audio. Instead of buying a bunch of new microphones, I bought the H-6, repurposed the old ones, and got great audio. With modern editing tools, you no longer need a clapper (those black-and-white movie things that say, "Scene 21, Take 4"). Your iOS device timestamps everything digitally, so Adobe Premiere Pro and Final Cut Pro will line everything up for you, automatically. It's amazing.

Recently, Zoom announced some new mixers that will plug directly into your iOS device—so you can add four or more lines of audio. Each line has its own gain and can be monitored with headphones. They're due out some time in 2017 and are sure to be useful. Check the *iOgrapher* blog for updates.

Using mixers is especially helpful when doing a multi-camera shoot. For example, say you're using one iPad for a wide-angle shot, another for a medium shot, and two more for over-the-shoulder shots. You would wind up with audio for each one of those shots. By adding a mixer, plugging in the MXL mixer into the two people filming, for example, you'll also have two lavalier mics going out to the two people talking.

In the end, you'll have one great audio track—the one that came from the two lavaliers. However, you'll also have the three other angles with reference audio, that didn't come from a microphone. They'll sound much worse, coming from the broad iOS mics. The mixer eliminates this problem, by allowing you to isolate the good audio. You can easily make a multi-camera edit and select the one good audio track in Premiere Pro or Final Cut Pro. All the angles will be lined up correctly, with that good audio from your main footage track. You'll wind up with everything you wanted.

When DSLRs first came out, you needed a mixer to capture audio, and then you needed a clapper board, or a guy clapping his hands. You needed a video guy and an audio guy. In postproduction, you had to use the claps and line up the tools as best you could. You'd get it eventually, but it took forever.

Now, your iOS devices and computer all have an Apple clock embedded in them, so they all sync. Everything is lined up automatically. It saves tedious work, and comes out perfectly. It's amazing to see how far we've come.

You can spend fifty dollars on a lavalier mic, then $229 on a four-channel mixer, attach both to your mobile device, and get professional-quality audio that can be fine-tuned to perfection. In the old days, you needed tons of thick, XLR-type cabling, plus a mixer that needed to be carried in on a car. They were huge. If you were doing feature-film-type stuff, you couldn't sync the audio on set, because big movie cameras only shoot film. They don't take audio in. Commercials, television, and movies all needed their audio to be done separately.

Now, you can do everything yourself, with an iPhone, a couple of mics, and a handheld mixer. People like Pewtie Pie, a YouTube star, can sit in front of their iPhone screens, upload to YouTube, get three million subscribers, and, boom, become rich.

Don't forget: audio is essential. Good video is nothing without great audio, and for the time being, great audio requires a mixer, when recording multiple voices.

PART THREE

ACCESSORIES

CHAPTER 6

CASES

||||||||||||||||||||||||||||||

It might seem weird that an ex football player, stockbroker, actor, and Hollywood producer created a case for iOS filmmaking, but only when you don't know the whole story.

When I was a producer, working with Mel Gibson's Icon Entertainment and the Lifetime TV Network, I saw how expensive the industry standard tools were. If you were a young kid starting out, you couldn't go out in your backyard and make a movie. It was too expensive and complicated. That bothered me.

When I moved on and taught media, I noticed almost every kid had an iPhone or iPad, and their cameras were nearly Hollywood quality. I had them get out into the world, to shoot and experiment. But there was a problem: the kids kept coming to me with movies that looked like *Cloverfield* or the *Blair Witch Project*—so shaky it literally made me sick.

The content was great, and so was the picture, but the shakiness ruined it. And the audio was crap.

They were such simple, fixable problems, but so vital. iOS devices had everything needed to make professional-quality films on the cheap, but they were missing two things: stability and the option to add accessories comfortably, like microphones and lights.

I could picture it: iOS devices needed some type of support system, with handles on both sides, for stable filming. It also needed to support the addition of microphones and lights on top. Finally, it needed to support additional lenses. If you wanted to get close up to get better audio with a shotgun mic, you needed a wide-angle lens. If you were filming a football game, you needed a telephoto lens, to get closer to the action.

I couldn't find a case like that on the market, so I made my own.

By adding stability and spots for accessories that improved

audio, lighting, and lens variety, the iOS went from great-but-flawed to having everything you needed to produce a professional-quality film. It was like having a Hollywood studio in your pockct. All it took was a little case that optimized everything Apple had already included.

At the time, I was an award-winning teacher who thought he'd be a teacher forever. I was mainly concerned with serving my students and getting the most out of limited budgets. Most schools had the older, full-sized iPads. They have great cameras but are too hard to use without a case. They were sitting there unused, wastefully. I made cases for all devices, to allow teachers (or whomever else) with old devices to repurpose them.

I wanted my students to have the ability to learn film-making by constant practice and play, but it had always been tough on limited budgets. I also wanted things to be easier for teachers. I wanted to make a difference in media education, but tight budgets were a reality I had to accept.

Most schools had good equipment that still worked—old tripods, iPads, microphones, and lights. The problem was that they needed a way to be repurposed. By creating cases for iOS devices, and adding some adapters, all of that old equipment could be used. It was cost-efficient for schools, simple for teachers, and fantastic for students.

It was also a timesaver for busy teachers and antsy students. Before iOgrapher, we had five cameras at our school: three DSLRs and two Canon broadcast cameras. They were awesome, but you had to lug around these big cameras, and transfer footage with an SD card, and they'd get lost all of the time. It was a huge time sink, and the more time wasted, the less the kids learn. I couldn't get as much work done. I needed more cameras, but even in a good school district, I wasn't getting more than one or two a year, at $3,000–$5,000 each. With 130 students, that wasn't nearly enough. Using iOS devices was far more realistic. Students started going out and filming on weekends. The world became their film set.

I was able to buy ten iPad minis, about $400 each, for the price of one big camera. They're far more versatile than the big cameras; they can be used not only for media and filmmaking but also for digital photography, animation, editing, streaming, and more.

Using iOS devices is a game changer, and the iOgrapher unlocked their power. I made sure it would be cost-effective and was able to keep the price tag at fifty-nine dollars despite producing locally, rather than in China. We also give educational discounts for schools. The iOgrapher is intuitive and simple to use; all you do is pop your iOS device into the case. The bottom of each case has a ¼-20 screw on it, which is the screw to attach any standard tripod.

To solve the problem of lenses, I looked back to the Sony Handycam and other old, handheld camcorders. Companies made extension lenses for those, so I thought, *Why not do the same for iOS devices?* The old ones were 37 millimeters, great, and inexpensive. We could use the same designs and molds they used to use, but update them for iOS devices. We used those 37-millimeter lens threads, and they came out great.

Before iOgrapher, nobody had made a case with handles on both sides. I couldn't understand it. I'm no genius, so it must have been a stroke of luck. Most cases had been designed to be as skinny and light as possible, which is fine, but why not also design a case that amplifies the power of these incredible tools? iOgrapher was counterintuitive to conventional accessory companies, but it was a game changer for creators.

I added handles on the sides for holding, a mount to the bottom for a tripod, and screws on top to mount lights and microphones. I made them the same, standard size as accessories that already existed. Then I added the options of wide-angle and telephoto lenses, and an adapter for DSLR lenses. It was a simple invention, but it connected

all of these underused resources, turning the powerful, yet flawed iOS devices into professional filmmaking tools.

I couldn't believe how quickly iOgrapher blew up. Right away, schools like Harvard and USC, professional sports teams like the Boston Celtics and Miami Dolphins, and TV stations like the BBC and CNN called me raving about our products. I guess I had been at the right place at the right time, with a product I personally wanted to use in my classroom. It turned out, though, that the uses would go far beyond schools.

Atop the case sit what are called *shoes*. A shoe is a space on which you can securely screw an accessory. Accessories like microphones and lights have a little adapter on the bottom, which screws into the shoe. There are hot shoes

and cold shoes. Hot shoes have electricity going through them, and cold shoes don't. On Canon DSLR cameras, for example, you plug your flash into a hot screw, and it talks to your camera so that, when you press the button, it knows to flash. Cold shoes are for secure holding of microphones, lighting, and other accessories.

Our iPhone cases have two shoes, plus another ¼-20 screw on the side, to which you can add a third shoe mount. A microphone and a light is generally all you need, meaning two is typically enough.

The iPad cases have three shoes on top, for a microphone and two lights, if necessary. Our newest products also have a ¼-20 screw on both sides, and our iPad Mini 4 and Pro cases have *two* screws on both sides. That's good if you're using an app that shoots vertical video (which I think is horrible, but it exists), so you can add cold shoe screws in there and have five cold shoes. Maybe you want a microphone and light on top, then two more lights on the side. You'd wind up with an incredibly well-lit subject, because you've got three lights just as they do in Hollywood, minus the expensive cost and difficult setup.

Microphones and lights are incredibly easy to set atop the iOgrapher. Most accessories have two mounts on them already, so you pop them on top, tighten, and you're good to go.

Some people are starting to put their GoPro cameras on top of the iOgrapher, too. That allows you to do as I explained earlier, and shoot with the GoPro's awesome camera, while using your iPhone or iPad as a monitor, via the GoPro app. Then, you can actually see what you're shooting, start and stop via the app, and wirelessly transfer your footage to your iOS device.

Shoes are just ports to attach *anything*, so the options are endless. That's the beauty of the iOgrapher.

If you're a beginner, don't get overwhelmed. You can use iOgrapher, on its own, for stability. The shoes are there because you have options to accessorize, but they aren't required. Add microphones, lights, GoPros, and so on if you need to, but more importantly, get out there and make great content.

I know I'm biased, but I wholeheartedly recommend the iOgrapher cases for *anyone* with an iOS device, at any level. Even if you're filming Junior's weekend soccer games, use an iOgrapher for stability. If you want to take a step up, buy a small monopod so you can be mobile, place it on the ground, and get nice, stable shots. You can walk around with it. It increases the video quality tremendously, takes up almost no room, and makes filming more comfortable.

iOgrapher supports the iPhone 5 and all subsequent versions, and every iPad from the beginning until modern day, except for the 12-inch iPad Pro because it's too big.

CHAPTER 7

TRIPODS AND MOUNTS

The iOgrapher's handles are perfect for when you're carrying your iOS device around, but there will be times when you'll need to mount it on a tripod. For example, you'll want to use a tripod when you need a stable shot, or you're filming sports, and you need to follow the action from left-to-right and want smooth movement.

When I was teaching broadcast journalism, we at least needed a monopod, if not a tripod. The beauty of a tripod is I could send students out on their own, and they could film a news report by themselves. They mount the camera, hit record, stand in front of it, and start talking. Tripods allow mobility and freedom.

Telemundo recently wanted to send a news reporter down to Mexico City to report on a breaking story, but they couldn't gather a crew in time. The reporter traveled alone, with her iPhone and a portable tripod, plugged in a handheld microphone, and did the story without a crew.

Like all accessories, tripods come in all shapes, sizes, and prices. You can get a cheap one on Amazon for fifteen dollars, and they go up to thousands of dollars for people who need to have the best of the best. As I explained earlier, all iOgrapher accessories have that ¼-20 screw on the bottom, on which all new tripods will fit. If you have an older tripod, you'll need an adapter.

The cheap, fifteen-dollar Amazon tripods will work fine in most cases, when you're starting out and filming yourself. If you're going to travel, do a lot of camera movement, or anything on a professional level, though, you'll want an upgrade.

If you're looking for an affordable, high-quality action tripod, we developed one here at iOgrapher that was released in November 2016. For eighty-nine dollars, this great, entry-level tripod allows you to fluidly pan left and right, and it even has the ability to turn the iPad vertically if you'd like. It comes with a travel bag and has a grip on it, and it opens to over five feet tall. It's more stable than other tripods on the market—we developed it to enable coaches and other users who need stable panning solutions to follow action shots. When you loosen other tripods, they flop around. Ours doesn't, and you can follow the action back and forth, left and right.

I've traveled the world with it in my backpack; it's simple to use, easy to set up, and highly durable. If you're on a tight budget, a cheap Amazon tripod is good enough, but if you can afford the eighty-nine dollars, our tripod is well worth it.

That said, action tripods require a steady hand. If you need especially smooth motion, whether left and right or up and down, you'll want a fluid-head tripod. They're used to film all professional sporting events, and they're used as well as in feature films, when cameramen want a nice, smooth action move of somebody walking left to right. A fluid-head tripod eliminates the jerkiness you'd have with an action tripod, but they're expensive. Manfrotto's entry-level one costs $250, but it's durable. If you're filming sports on a serious level, you'll need one.

Our tripod is great, but it isn't to be confused with an expensive one used in movies. It's not as smooth, but for people who can't afford Hollywood equipment, it is more than sufficient.

If you need more support, you may want a shoulder mount. The two iOgrapher handles are often sufficient, but if you're doing a long scene or moving around at lots of different angles, a shoulder mount will help you with comfort and stability. They adjust for height, and will fit nicely on your shoulder, even if you're a big guy like me. A shoulder mount gives you the ability to move around and see the screen in front of you. You can also add more accessories on top, because you have a bigger support base. Our shoulder is mount is light, made of aluminum, and sells for $120. If you're doing some long, steady shots, it's a great addition to your toolbox. Prices extend to the thousands of dollars, but any shoulder mount works fine for an iOS device, and they all have ¼-20 screws on them.

Monopods are also great for mobility and stability. They

start at twenty-five dollars on Amazon, and we sell a quality Manfrotto one. I always use one when filming conventions. Recently, I was at the consumer electronics show in Las Vegas with my sales employee, and we went out and filmed iOgrapher users and products to show on iOgrapher TV. A monopod allows me to rest the iOgrapher, so I don't have to carry it all the time while filming.

To add another element of portability, you can use Dinkum Labs' monopod mount. It looks like a monopod that has a bendy torso on it, but the bottom has a big clamp you can attach to a table, railing, tree branch, or any surface. It's great for when you don't want to carry a tripod, but you know you'll have something to clamp to. It's a portable, inexpensive tripod replacement.

It's called the Action Pod Pro, and it costs forty-five dollars. It can hold a fully loaded DSLR camera, so it certainly supports the light, portable iOgrapher, and your iOS device, regardless of how many accessories you attach. It's a great little mount, and I take my three Action Pod Pros with me wherever I go, because I never know when I'll need to clamp to something.

When I'm filming tutorials, I'll get a wide shot of me talking to the camera, then clamp one of these onto a desk with an iPad, to get a close-up shot of whatever I'm teaching. For example, it will get a close-up shot of me showing how to screw on a lens. It's incredibly simple to set up.

Another thing you may want is a slider. A slider fits on top of a tripod, or even a table, and allows you to move your camera from left-to-right or front to back. When you see a movie shot where someone is walking and the camera slides along with them, they either used a slider or had someone running across a big track with a slider car. Since we're not in Hollywood, the cheap substitute is buying a small, sub-$100 slider. Of course, they range into the thousands of dollars, too, but an inexpensive one from B&H Photo (a big NYC camera store with a great online store) will do. Schools often buy from them. Search for "video camera sliders."

I personally use the Eddlkrone Slider Plus. It attaches on top of my tripod and slides easily. It's hard to explain, but the design makes sliding much smoother than others. That's why it costs $500, but if you're going to use sliding shots a lot, it's worth it. If not, the sub-$100 ones will do.

If you're going to be doing shots where you start out really high then come in low, vice versa, or are swinging left to right, you may want a Kessler Pocket Jib (go to KesslerCrane.com). It costs about $1,000, but it's great for these advanced shots, giving you a fluid, professional up-and-down fluidity.

Adding a jib and sliders is only necessary for serious filmmaking, but it will make you look like a Hollywood filmmaker with thousands of dollars of equipment for a fraction of the price.

CHAPTER 8

LIGHTING

Most camera apps have a way of adjusting apertures and lighting settings on their own, but oftentimes, that won't be enough. You're going to need some lighting. The other day, we were filming in a well-lit kitchen, shooting a cooking video for a client. She had beautiful, blue eyes, but we couldn't get them to show, even in the luminous kitchen. We had to add some LED lighting. Once we did, her eyes popped, and we nailed the shot.

Another example would be when you're shooting an interview, but you aren't out in perfect sunlight. You would need a little LED light to sit atop your iOgrapher. We sell some great ones, and most have filters to change the color temperatures or brightness. If you're in a low-light situation, or it's dusk or nighttime, lighting is essential.

Basically, the only time you *won't* need lighting is if you're in perfect daylight, outside, at the beach, or shooting little, fun stuff that doesn't need to be perfect. Then you can bring the aperture and exposure levels down a bit in editing, and rely on nature's big light. But the world isn't perfectly lit most of the time, and you're going to need to fix that.

I don't recommend getting cheap LED lights. I've bought plenty of twenty-dollar LED lights on Amazon, but they carry twenty batteries in them, and they're gigantic and heavy as hell. They break a few days later. You can use them for small stuff, but they won't adjust more than a slight dim, and investing in quality lights will save you money in the long term.

The granddaddy of them all is a ring light called the Roto Light Neo. It costs $450, but it sits atop your iOgrapher, it's light, and it runs off AA batteries or plugs into an outlet. It's programmable; you can adjust light colors significantly. Movie studios typically use tungsten light, and the Roto Light Neo allows you to adjust to match their lighting. Another example would be if you were inside a mall and wanted to use fluorescent light; the Roto Light Neo would allow you to match the color temperature of the light. That's crucial, because if you have varying color temperature lights, your image will look strange. Everyone would notice, but you can fine-tune the problems away with this quality light.

The Roto Light Neo also has some fun special effects. You can add a blue filter, put it outside your window late in the afternoon, and simulate lightning. Your movie will look as if there was lightning going on in the background. It's high quality, customizable, fun, portable, and well worth the money if you can afford it. It should last you years. It has a shoe adapter for accessories, as well as a ¼-20 thread, so you could technically attach it to a separate tripod and angle your light in different directions.

Typically, the Roto Light Neo will be all you need. Still, many people like to have at least three lights in their toolkit, so that they can have the background lit, the filming subject lit, and then play around with a third, according to the situation. That's unnecessary for interviews and simple filming, but having three lights is helpful, if you can afford it.

If you're looking for something more affordable than the $450 Roto Light, look to our friends at LumeCube. They're a fellow Kickstarter success story and are rocking it in the light world. The LumeCubes are compact, waterproof cube lights that cost under $80. By using their LumeCube app, you can program them to be used as flash for photography. They have several levels of brightness, and they are small enough that you can add two or three of them to your kit and mount each of them to an iOgrapher for great lighting in interviews. The company has plans to develop more accessories, such as barn doors (to diffuse lights and focus more on one subject) and filters, to name just two. You can even mount LumeCubes on their own tripods and place them around the room for nice lighting.

Here's how you would create a whole, professional set, in terms of lighting. Check out Rotolight Neo's three-light kit, which will run you about $1,400. It gives you three lights, three tripods, three A/C adapters, some filters, and an awesome, durable carrying case. I've used it for interviews and cooking shows; it's all you need for most indoor situations. Those three little lights are powerful enough to light up a big space. You can put them anywhere, because they all fit on a tripod or your iOgrapher. They're only necessary for serious filmmakers, but they're a great deal that will get you great lighting.

If you don't want to spend that kind of money, stick with the LumeCube lights. Many of them have ¼-20 screws, too, so you can mount them on your tripod or iOgrapher and move them around. Screw one of the lights on top, mount it, and it will light up the person you're interviewing, right in front of you. Our iOgrapher Flex Tripod is perfect for this—it's a mini tripod you can bend and tie around a pole, or anything else, for under thirty dollars. For well under $200, you'll have great lighting options.

If you're on a serious budget, there are cheaper options. Go to any local hardware store, like Home Depot, and get some work lights. They cost a few bucks, and you can use wax paper in front of the lights as diffusers. They don't have dimmers or any adjustment settings, so you've got to place them properly so that they don't overpower your subject. Regardless, using work lights is an age-old tradition for people on a budget that still works today.

Don't forget lighting; it's essential. Even though iOS devices have great cameras, you're going to get some grain in your video if lighting isn't perfect. By spending a few dollars to get quality lights, your video quality will improve massively, and you'll be happy in the long run. You might think the iPhone 6s Plus's 4K camera is invincible, but it only will be if you give it the information it needs, through quality lighting.

CHAPTER 9

LENSES

World-famous photographer Annie Leibovitz, who photographed John Lennon on the day he was assassinated, has been quoted as saying, "The iPhone is the snapshot camera of today." iOS devices have amazing lenses. That said, they have a digital zoom, which locks you into one range. When you pinch and zoom in, you lose some video quality, because the video is getting digitized. To alleviate this problem, I wanted an optical zoom, which uses glass to make things closer, similar to how eyeglasses work. When you have bad eyesight and put different lenses on, you see different things. I wanted that lens ability for my iOS devices, and that became one of the most important features of the iOgrapher.

Apple's recent release of the iPhone 7 Plus was a game changer, though, with its dual-lens system. It has a great, wide-angle lens, in addition to a 10x-telephoto lens that will allow you to create a depth-of-field look when shooting photos and some video. At the time of this writing, lens makers haven't developed a lens that works with this new feature, but they surely will soon.

You may also have lens problems when live streaming with Periscope, Facebook Live, or YouTube Live. Often you'll be using the "selfie" camera lens that is facing you, so you can see your phone's screen to answer comments. It's a close-up lens, though, and looks strange to the viewer—it's as if they're standing right in your face. There are little clip-on, wide-angle lenses you can buy on Amazon for ten dollars that will show more of what's going on around you as you're live streaming, and spare the reader that weird, close-up feeling.

I've always wanted to use my wide-angle lens for 99 percent of filming, and now I can. It gives me the most possible viewing space, so during editing, I have more options to crop and change things around. I shoot everything in 4K, but upload to YouTube in 1080p, and that extra quality allows me to zoom in and do tons of cropping and fun stuff that doesn't degrade the image, because it was shot at such high revolution. Shooting with a wide-angle lens gives you the biggest possible canvas to play with in editing.

Our iOgrapher wide lens isn't *just* a wide-angle lens, though. You can screw it apart, and it will become a macro lens, which allows you to get really close to subjects, such as for filming ants building an anthill. The shots are amazing. People don't understand how a lens could be wide and macro; it's because the lens has two pieces. It's perfect for your iOS device.

A photographer for *National Geographic*, based in Okinawa, uses all of our products and loves his macro lens, which he uses to get gorgeous shots of bugs.

Compared to the standard, iOS lens, the iOgrapher lens gives you significantly more viewing space. You're going from five or six feet wide to ten. Having a canvas of double the size gives you so many more options in editing. I'm a big believer in getting the most that you can, so you'll be happy, later, in postproduction. Options are never a bad thing.

We also have a 2x-telephoto lens, which was requested constantly by sports coaches and wildlife photographers and videographers. Let's say you don't want to get too close to that grizzly bear. The 2x-telephoto lens makes the shot look twice as close as you really are. It's infinitely better than pinching and using the digital zoom, which would degrade video quality. The 2x-telephoto lens lets you get *optically* closer, which is crucial.

It's also perfect for sporting events, when you need to zoom in. Make sure you have a steady tripod for the telephoto lens, though, because the more you're zoomed in, the more any sudden motions will show up on video. Tiny movements up and down or left and right will show up gigantic, because you've zoomed in so much, to get close to the action. We're also beta testing a ridiculous 10x-telephoto lens, at the time of this writing. That would be exciting to use for very close-up filming.

My all-time favorite lens is the Moondog Labs Anamorphic Lens. Growing up as a big movie buff, I considered *Laurence of Arabia* to be one of my favorite movies. Big epics like that were shot with what's called an *anamorphic* look, which is super wide, cropped on the top and bottom, which leaves you with an amazing landscape. It's the same lens Quentin Tarantino used recently to shoot *The Hateful Eight*.

The wider your shots get with a wide-angle lens, the more bending you get on the fringes. For example, a fish-eye lens shoots superwide video, but everything looks bent.

Anamorphic lenses give you the width, but there's no bending. It's awesome.

Anamorphic lenses aren't that common, but that's only because they haven't been in full supply for iOS devices until recently. Moondog Labs was started by a couple of ex-Kodak engineers who started making their own lenses after Kodak sadly went down the tubes. They launched on Kickstarter and are growing rapidly, for good reason. They make great lenses, specifically the one for our case that has a 37-millimeter thread on it.

As we discussed earlier, the recent Sundance film *Tangerine* was shot entirely on an iPhone 5. They used an anamorphic lens, which gave it the filmic look they were going for. It was made for pennies, got a distribution deal, and is on iTunes.

You can also mount DSLR lenses, such as a Canon or Nikon, to your iOgrapher device, by using an adapter like the Turnikit 37-millimeter adapter. The adapter fits on your case, and the lens fits on the adapter. It's amazing if you're looking for that soft focus, depth-of-field look that films have. Turnikit first launched on Kickstarter, too, they manufacture in the US, and their founder is a wonderful guy who loves his product as much as I do.

The adapter is perfect if you already have an expensive DSLR lens that you want to repurpose and get the most out of. It's also great for super-close-up shots of people, if you want the person to be totally in focus, with everything else out of focus. You'll need great lighting, because those lenses soak up light like crazy, but you'll get a nice, artistic, soft focus. It won't be like the supersharp focus you'd get on your Nikon camera itself.

You don't need to get fancy, unless you're a professional filmmaker, but with entry-level accessories like our wide-angle, macro lens, you can turn your iOS device into a professional filmmaking kit. Take the *New York Times'* word for it. They wrote, "An iPhone case ready for Hollywood. iOgrapher helps improve your smartphone video." They know what they're talking about.

PART FOUR

APPS

CHAPTER 10

APPS FOR FILMMAKING

As a longtime filmmaker and content maker, I'm fascinated by the fact that I can do *everything* on one device, by filming with iOS devices. It's so much more convenient than traditional filmmaking. The options are endless once you start looking into all of the cool apps out there, but you can even start with what comes on your iOS device. You can get started with an iPhone or iPad and the free iMovie app that is already on your phone. It's powerful.

Back in my Hollywood days decades ago, I used to report for different magazines, such as *Post*, the postproduction world's magazine. As a reporter, I would get in free to conventions (it was a way for me to get some great, free equipment for filming), but I needed a big camera, giant tripod, and all kinds of cables for audio.

One year, pre-iOgrapher, I brought nothing but my iPad, as an experiment. I felt naked, but I wanted to try it. I got high-quality interviews on the fly, and I was amazed. All I had was a side satchel (man purse, if you prefer). It was so mobile. I sat down right there and edited in iMovie.

The quality was about as good as the previous year, when I had lugged around bags of equipment. I made a few rookie mistakes, though. I learned the hard way that iOS cameras are auto-focus, constantly looking for focal points, pulsing in and out of focus constantly. As we've discussed, that's easily fixed by locking your focus. The audio was solid, and from that point on, we were off to the races. Ending up with poor quality that first time was annoying, but I'm glad I've gone through all of the failures so you don't have to. That's why I'm writing this book.

Remember: Always lock your focus! (See **bit.ly/lockfocus** for detailed instructions.)

Once you start downloading apps, your options multiply exponentially. You can edit photographs, add colors, and fix things with apps like Photoshop. You can draw things. When the idea for iOgrapher popped into my head, I whipped out my iPad, opened the Paper app, and started sketching a design. I'm not a good drawer, but it was good enough to show my high -school students and see if they thought it was cool and might use it. iOS devices are the Swiss Army knives of creativity. Everybody knows you can do business activities on them, but that's a little boring, don't you think?

The first few generations had decent, high-definition cameras. They were good enough. Now, our iPhones and the new iPad Pro can shoot in 4K. We have apps that can trick out your shots in high quality, color-correcting on the go, and sharing instantly with anyone on social media. There are apps that will do your titling for you, or help you storyboard your movie. There are apps that capture audio only. If you can think of a story to tell, your iOS device is literally the only tool you need to tell and share that story. That's why I love mobile filmmaking so much; it levels the playing field so that the best stories win, rather than the best equipment and most money.

Before we get into apps, though, realize that this is a fluid, fickle world. Apps often die the instant marketing money runs out or a poor update is released. I'll only mention the few I expect to last long-term, but our website will be updated with our latest, recommended apps (iOgrapher.com/apps), and you can always shoot me an email at davesbook@iographer.com with app questions.

iMovie is backed by Apple and already on your phone, so it's quite safe and the perfect place to start. It's your bread-and-butter editing app, and it's free. It's simplistic and intuitive; you can trim your footage by dragging bars left and right. It reads directly from your camera roll and finds everything you need to be imported into the app. You can even shoot *with* iMovie. If you're editing and say, "Oh my god, I need this bit of footage of that flower outside," you can insert that clip right then and there, within iMovie.

Open the iMovie app, and you'll see a tab called "Projects," which will display your existing projects and allow you to create new ones. Your projects turn into movies, or even trailers. My son used to love making silly, Hollywood-esque trailers; it was ridiculous, but great. There's also a "Videos" tab, with your camera's clips, and "Theater," with your completed films. When in theater mode, you can share and play your movies via Apple TV.

When you create a new project, you're presented with template options. One of my favorites is the CNN iReport template. If you're a journalist, or an aspiring one, you can get out there and film stories, and when you upload with that template, you may be picked from CNN to get on their website or even on TV.

Once you select a template, you import footage—audio, video, photos. You can crop. You can change audio levels. You can rotate footage, or manipulate it any way you want to. You can record, to add shots. You can add audio by itself. You can do picture-in-picture. The browser sorts your video, photos, and audios separately, so it's easy to navigate. You can do any type of basic editing, right on your iOS device.

You can add basic filters to change the coloring of your video, with options like black and white or blockbuster. You can make your film look vintage, or like *The Matrix*. You can add theme music and transitions, like fade-ins and fade-outs. Once you're done editing, you can add titles and export your footage to share it with the world, wherever you'd like.

One of the best parts is the speed at which you can produce videos. When I first started reporting from trade shows with my iPad, I could edit at the show, while eating lunch, and be done hours before everyone else. They needed to go back to their office, upload the footage to a computer, and edit there. I was already on to the next story.

These are just the basics, but we have full tutorials on *iOg-rapher Academy* (**bit.ly/iogacademy**) iMovie is a simple app, but it's perfect if you're starting out. My intern recently edited his first little piece of content in iMovie. Soon enough, I'll introduce him to Final Cut Pro, because as great as iMovie is for beginners, it has limitations.

iMovie's biggest deficit is its lack of titling options. You can't add titles on top of footage; there are no layers on them. I expect an app to fix that one day, but for now, say you have great interview footage, but you want B-roll on top of it. You're interviewing a guy about his invention, and you want to show someone using it. To do that you'd need *layers*, but iMovie only allows for one layer. Unsurprisingly, this is the top pet peeve of news organizations. iMovie also has a dearth of title types—of template, music, and so on. They're simple, cookie-cutter, and look the same after a while.

My go-to iOS app for filming video is FiLMiC Pro. It's your iPhone or iPad camera on steroids; it increases the number of megabits per second you can record in (from twenty-four megabits per second to fifty), giving you a clearer picture and more options for postproduction later. Whenever you see something professionally done with an iOS camera (like *Tangerine* or the Bentley commercial mentioned earlier), it was filmed using the FiLMiC Pro app. It gives you higher resolution, as well as control over your exposure levels (control of the amount of light that enters your shot) and frame rates. You can do white-balancing and lock your focus. It has a digital zoom, but it's much better than the default iPhone one, because it smoothly follows your subject and locks your focus. The resolution settings can be adjusted for old TV mode (4:3), today's 16:9, or 1:1 for Instagram. You can change frame rates from film-like twenty-four frames per second, all the way to 240 for slow motion, and there's also a time-lapse setting. You can set presets for different locations, such as for my garage "studio" where I have FiLMiC Pro presets for my green screen. I hit "garage," and I'm ready to shoot.

FiLMiC Pro now has a remote app for your iPhone, too, called FiLMiC Remote. If you're using your iPad to film, you can control it through your iPhone via the app. You can put the iPad on a tripod in front of you, step back, bring out the phone, and connect it. Remember to lock your focus, then feel free to mess around with the exposure and frame rates, and whatever else you'd like. You can press "start" and "stop" on the remote, so filming yourself is easy.

The app also has a feature that allows you to start and stop filming using your Apple Watch. If you're filming a news story, for example, you can set up your shot, stand in front of the camera, and check your Apple Watch to see if it's framed correctly, then hit a button to start shooting. You no longer need someone with you to help.

Another crucial aspect is that FiLMiC Pro supports external hardware (accessories). For example, if you attach the Moondog Labs Anamorphic Lens on your iOgrapher, it'll come out squeezed and long. Your iOS camera isn't programmed to use that type of lens, but FiLMiC Pro has an adapter setting that will make the shot look beautiful, as intended. It supports 35-millimeter Nikon or Canon DSLR lenses, by flipping the image so that it shows up properly, a necessity.

The biggest selling point, though, is that it's one of the few apps that lets you monitor live audio. You can change the gain, and you'll never have the problem of realizing in postproduction that your microphone wasn't on, or you had some fuzzy sound interrupting your audio.

The app is simple to use, costs eight dollars, and is always up to date; you shouldn't film without it. Make sure you visit *iOgrapher Academy* for a full tutorial on FiLMiC Pro (bit.ly/iogacademy).

I'm a big user of Adobe Creative Cloud, Adobe's suite of editing tools. I was an Adobe educational leader as a teacher, so I've been using their applications for fifteen years. They're great for students, and they all coordinate with each other. You can start by bringing your footage into Adobe Premiere Pro (their nonlinear editor), right-click on a piece of footage, and send it to Adobe After Effects if the shot needs further editing (like a green-screen shot). After Effects is a professional-level compositing tool, and it's a click away. So is Adobe Audition, where you can edit audio.

Adobe Premiere Clip is the basic video editing tool you can use on your iOS device. If you're willing and able to pay the fifty dollars a month for an Adobe Creative Cloud membership, you'll get upgraded access to Premiere Pro and all the other Adobe editing tools. It's a step up from iMovie or Premiere Clip and a great investment if you can afford it and are doing more serious filmmaking. Set up your free Adobe Creative Cloud ID, buy a membership, and log in. Adobe stores everything in the cloud, so you can move files from device to device, or from a cloud app such as Dropbox to your device, as well as from Adobe app to Adobe app. Moving within Adobe is awesome, such as if you're editing photos in Adobe Lightroom and want to move them to Premiere Pro for a film.

The range of Adobe's apps is remarkable. For example, there's one called Adobe Capture, an advanced color-editing app. Let's say you visited Hawaii and want to encapsulate its beauty. Take a photo, save it to your Adobe Creative Cloud account, and then you can extract that beautiful color scheme and apply it to other photos and videos you took there. Every photo and video you took can now have that coloring.

Another advantage is that you can do a quick, basic edit in Premiere Clip on your iOS device, but because that quick edit is synced to the Creative Cloud, you can access it later on your computer for more advanced editing. You can do your first cut in Adobe Premiere Clip, and then do a full edit in other Adobe apps, such as Premiere Pro (advanced film editing), After Effects (video), Audition (audio), and Photoshop (titling).

Adobe Premiere Clip is similar to iMovie, in that it's a free app that is great for basic editing, but has limitations in titling and a lack of advanced features. If you pay for the reasonably priced Creative Cloud account, you can get the mobility and flexibility of Premiere Clip on your first edit, and the advanced capabilities of the suite of Adobe apps afterward, to clean things up, add titles, and look professional. If you're using Adobe Premiere Clip to start, you'll have a big head start on people using DSLRs and fancy equipment, because they can't make that first edit on their iOS device.

If you pay the fifty dollars a month, you can do a basic edit in Premiere Clip on your phone, then access your footage on the Creative Cloud and import to Premiere Pro and the other apps for advanced editing. If you're willing and able to pay, the benefits are worth it.

Apple does have similar functionality, though, by pairing the iMovie app on your iOS device with Final Cut Pro ($299). You can do a first edit on your iOS device in iMovie, export it via a USB cable in an XML file, then do your advanced editing in Final Cut. Adobe is faster, though, because it sends your file to the cloud. Therefore, if you're working with a team, your editor can access the file as soon as you're done shooting, via the cloud. The editor can log in and edit immediately. If you have iMovie, you need to physically connect your iPad or iPhone to your computer to start working on the file. Besides that, the two programs are comparable.

Now, let's examine the suite of Adobe apps in more depth. Adobe Capture, for example, lets you capture the "look" of a photo. So, you can take a shot of the summer sky, bring that "look" into Premiere Clip, lay it on your video, and export that to Premiere Pro later for more advanced editing. There's also Adobe After Effects, a desktop tool for visual effects and compositing used by professionals on big, Hollywood movies. After Effects is the tool of choice of the director of *Godzilla*, Gareth Edwards. Next, there's the audio app, Adobe Audition, which you can use to fix and adjust audio and add soundtracks, among other things. Apple has its own version of these tools, but I don't personally use them, so I can't comment. I'm a big Adobe fanboy.

In the old days, you needed to spend over a thousand bucks to get the Adobe Premiere and After Effects combination. Then they would upgrade the software every eighteen months, and you'd have to spend another $700 to upgrade. Fortunately, that business model has been replaced by the Adobe Creative Cloud subscription membership. Now you can pay fifty dollars a month and get all new upgrades immediately, for free. It's far cheaper, and you're always up to date. You're leasing the software, technically, but you have access whenever you need it. You can also cancel or suspend your membership whenever you'd like, so you can pay for access while working on a big project, then suspend it while you wait for the next one.

Even with just FiLMiC Pro and either iMovie or Adobe Premiere Clip, you have all the tools you need to make great content. Other apps are great to look into for specific purposes. Really, though, just pay eight dollars for FiLMiC Pro, open up your iPhone or iPad, and start making videos. That's all you need to win Sundance.

CHAPTER 11

LIVE VIDEO APPS

||||||||||||||||||||||||||

Live video is changing the world. It allows anyone to host his or her own television show. I realized this when I was on family vacation in Hawaii, and my eight-year-old son was hot and got bored walking around Pearl Harbor. He wanted to go back to the hotel and swim, as eight-year-olds do. My wife took him back, and I was alone in this amazing place, Pearl Harbor. I love World War II history and have family members who fought in that war, so I went to the battleship *Missouri*. That's where they signed the surrender of the Japanese Army back in 1945, and I was in awe, walking around the gigantic deck.

I whipped out my iPhone, put it in my iOgrapher, added a microphone, turned on Periscope (which is linked to

Twitter), and all of a sudden I was connected to my sizable Twitter following. *Hey, I'm out here on the Missouri. Would anyone like to watch?*

Minutes later, 457 people were watching my live stream and asking me questions as I navigated the ship and narrated. Almost all of them were strangers, and it was remarkable to share the experience of showing them a place most of us had never been before. It rattled my worldview, giving me a visceral realization of the power of live video. We're still in the early days of live video; it's no gimmick, and it's growing rapidly.

Recently, I was sampling new software and hopped on Facebook Live to get some feedback and see if everything was working correctly. Sixty people watched, and we had a mutually beneficial conversation about my products, how to use them, and which apps and microphones I recommended. Those sixty live viewers turned into three thousand eventual viewers of the saved video.

Your options are endless. I have a musician friend named Rob Gonzalez, an amazing pianist and singer of jazz and other genres. We're working together to live stream performances to his fans, by connecting audio capture tools to his iOS devices. Without the tools, the music sounds tinny. With them, though, he can perform live from his living room, with just his iPhone. Fans are thrilled. U2 recently used Periscope on tour; Bono held his iPhone, walked around, and chatted with fans with perfect audio. Live video is an incredible way for anyone to connect with fans and build an audience—fans feel as if they're with you in person, in real time.

Live video is great for general business marketing, too. For example, a small-town, old-time bicycle maker could teach fans how to refurbish a specific type of bike. He could post archive footage to YouTube or Facebook and start building an audience. According to Facebook, video is going to be 95 percent of your feed within the next five years. The time to get into live video is *now*.

After the most recent NBA Finals, the NBA broadcasted "NBA Live" on Facebook. For TV viewers, the broadcast would cut out after an hour or so; the network had other shows to play. On the Internet, however, you could watch postgame stuff all night long. You could comment and talk with like-minded people. Live video is as great for audiences as it is for creators.

Facebook Live was initially restricted to the Kardashian-type celebrities with massive audiences, but Facebook pivoted, realizing it should be accessible to the everyman. Still, most people don't know how to use it. Don't just turn on your camera and walk around. The objective of video is *always* to tell a story. Occasionally you can get away with laziness, if you're streaming a football game or someone else's content, but otherwise, I want to hear your commentary on what you're showing me. We all want stories, regardless of the format.

Add a microphone, comment, and narrate your story. A lot of people show video, but you can't fully understand or hear what's going on. The key to Facebook Live is to tell your audience what's going on.

Live video, of course, is raw; you can't edit. That means you either have to plan beforehand or do something impromptu that will actually be interesting. It depends on your goals. I've created shows through live video that were all planned out, and I've riffed on the spot. I've included external videos, used green screens, and even included commercials. Planned and unplanned can both work, depending on what's best for you, your business, and your situation.

The Swiss Army knife of live video apps is Wirecast, made by Telestream. It's expensive ($495), but it's a great value if you can afford it. It gives you access to the Wirecast Cam iOS app, which allows you to use your iOS devices as cameras, switching between feeds as you like. They all sync to your desktop via the app, so you could place four or five iOS devices wherever you want, using them as cameras, and have separate audio feeds coming in from each. You can also connect your laptop to a soundboard, if you're at an event. It has instant-replay ability. It allows you to create a professional-quality, multi-camera TV show, live, on the fly. All of your devices sync via Wi-Fi and the Wirecast Cam app for your iOS device. You can stream to YouTube, Facebook Live, Twitch, The Cube, ChurchStreaming.tv, Wowza, and other outlets.

Recently, my sales manager and I streamed a Facebook Live of a CUE Student Filmmaker Awards Ceremony. I brought my laptop and three iPads in my backpack, connected to the soundboard for awesome audio, and we were professionally streaming without the need for bulky, cumbersome cables and devices. Wirecast allows you to stream directly to Facebook Live or YouTube Live. You can also record to your device, preserving all of the angles, and bringing in Skype calls for interviews. Wirecast is a professional tool for multi-camera, live experiences. The only cons are that it's iOS-only, and you need a laptop to operate it.

The budget option is the Wirecast Go iOS app, which is free to download. It doesn't allow multiple cameras, but it will upload video to your YouTube Live channel with the push of a button. It isn't yet compatible with Facebook Live. If you upgrade the app for $5.99, you can add logos, watermarks, or pictures to your stream. Of course, it's all compatible with microphones, lights, lenses, and your typical setup.

Regardless of which app you choose, both are great value, and we have full, zero-to-hero training on *iOgrapher Academy*, for free with this book (bit.ly/iogacademy).

Similar to Wirecast Go, Switcher Studio is a free app that allows people to film or live stream for up to twenty minutes a day. However, Switcher Studio works with multi-camera workflows. One device acts as your switcher, and you can add up to three other devices, giving you up to four different angles. They all connect via Wi-Fi or LTE. You can switch between views with the push of a button on your switcher, and you can stream directly to Facebook Live. Annoyingly, Facebook Live only allows you to shoot vertically, but by using Switcher, you can stream horizontally. That gives you a better view and more options.

Switcher Studio allows you to bring in photos and other graphic elements and to create your own titling. You can also connect the iPad acting as your switcher to a big screen at a convention, via an HDMI cable, so everyone

> Switcher Studio has released a new app, Switcher Go, which allows you to add presentations to your multi-camera shoots. This is great if you're live streaming and want to cut to a video or picture you shot earlier—maybe you're talking about an interview you did, and then you want to show a clip, live, before cutting back to you. You can download it now in the iOS App Store.

can see your multi-camera video. It's a great entry-level app, especially for students, because the teacher can use his or her iPad as the switcher, and students can connect their iPhones. I've personally filmed football games with one wide-angle camera on top, two iPad Airs running along the sidelines, and one more for tighter shots in the middle. I used Switcher Studio to change angles, and streamed live to a high-school sports website called The Cube.

There's an advanced version for twenty-five dollars a month that allows you to shoot more than twenty minutes a day and gets rid of watermarks. It can also connect to a desktop so that you can broadcast your computer screen, which would be great for teachers wanting to connect to a whiteboard.

Wirecast has a lot more bells and whistles than Switcher Studio, but it's more expensive and has a much deeper learning curve. Your choice will depend on your personal situation. I shot twenty-seven videos to teach Wirecast to iOgrapher students, as opposed to four for Switcher Studio. Switcher Studio is great for basic, multi-angle shoots. Wirecast is great for more advanced shoots, when you need things like multiple-depth pops or live graphics on the fly. Both will give you professional, multi-camera streams. Check out **bit.ly/iogacademy** for complete tutorials.

The fact that Facebook Live has opened itself up to everyone is a game changer. You can open the Facebook app, press the "live" button, and start streaming. The only downside is that you have to shoot vertically, but that does allow you to see comments underneath you video, so you can respond. You can use Switcher Studio to stream horizontally, but you'll need someone else to read you the comments. Regardless of how you do it, just start talking and connect with your audience on demand. It's amazing.

For our company, and most companies, Facebook is a huge driver of traffic. All of our fans are on Facebook, so Facebook Live has become the place to go to engage with fans. There are tons of other players on the market, some of which we'll discuss, but everyone keeps coming back to Facebook, because that's where we all are. It has the largest, most engaged community, with 1.71 billion active users sharing with each other.

Facebook Live has so many uses, but my favorite is education, of all types. As a business owner, I teach my customers how to use their iOS devices to their fullest potential, regardless of who they are. Our biggest audience is schools, but we extend to professional sports, music, journalism, and tons of other arenas.

I'm launching a show where I interview someone live on Skype, every week. For example, I'm going to interview the head of communications at the University of Missouri, who's going to discuss how he uses iPads and iOgraphers in the classroom. This should open a dialogue to get schools interested, while educating people in general. For businesses, the more you educate your fans, the more apt they are to buy. Educating customers also builds trust in your brand, by showing them the value you provide.

One of the biggest benefits of live video is that viewers are far more engaged than while watching non-live videos. People will respond to you in real time, rather than loosely considering commenting later. It's a great way to connect; they hear, "Hey, I'm from Minnesota," and immediately react. "Wow. My wife is from there. I love Stillwater!" It starts a conversation, because you don't have to wait days, or lifetimes, for a response. It's more immediate, and more human.

The sports teams we work with love live video, too. I've seen NFL fans chat with their quarterbacks after practice. The quarterback hops on a live stream, and fans ask him about next week's game. When he acknowledges their presence, they go nuts. It's a new world, allowing fans to talk with their favorite players. People love their favorite player mentioning their names. That's a great practice, to call people by name in live video.

Another example would be for app makers. Plug your iPad into your computer, and use Wirecast to share your app's screen. You can then show people how to use your app. The implications for learning via live video are endless. It's not about showing people parties or ranting about something. You could be a travel agent doing a show at the Parthenon, encouraging people to book a trip to Greece.

One of the other big live video players is Periscope. Periscope blew up alongside Meerkat at SXSW in 2015, as the first live-streaming tools that simply required a click to start. I used to like Meerkat better, but it was recently shut down, and Periscope is connected to your Twitter account. That makes engagement significantly higher. It's a simple app where you type in what you're talking about, and it goes out to your Twitter following and the world. People can then comment, and you can get into conversations like I did on the deck of the battleship *Missouri*.

There's a big politician in Canada who uses Periscope constantly. It keeps spreading: to politics, sports, education—anywhere that people want to build fans, constituencies, and communities. Live video gives the viewer a voyeuristic thrill, and it allows them to interact with people they may never meet in person. It gives regular people a voice; maybe your hero will hear your idea and like it.

I personally use both Facebook Live and Periscope, but which one you use will depend on where your community hangs out. As a business owner, I need to be everywhere on social media. The bad part about Periscope is you can't schedule live streams, so the only way to let people know is by telling them in a Tweet. On Facebook, you can schedule streams and build an event on your page so people can sign up and receive alerts. Periscope has launched Periscope Producer, which enables you to broadcast several sources of video on the fly, for multi-camera streams.

The biggest drawback of live video, in general, is that people can say anything they want. You have to learn how to deal with trolls, especially on Periscope, where nonfollowers can discover your stream. They don't know you, so they'll start swearing at you without even listening. You can block them with a quick hand motion, but it can be frustrating. On Facebook, only people who liked your page can participate, so it's not as bad. Things also tend to get lost quicker in the Twittersphere, versus Facebook. Periscope has its place, though; Kevin Hart jumped on and riffed for ten minutes recently—he was funny, and got millions of viewers. It's all random, and it all depends on the situation.

If you're a marketer, business owner, educator, or anyone trying to connect with an audience, live video is the way of the future. You'll be a dinosaur if you don't get involved. It's not about being famous or vain; it's about telling stories to people looking for them. It's about getting your product or service's message across. Live video is the easiest, most cost-effective way to build your brand and put a face to it. It's how you build engagement and trust, both of which lead to sales and happy customers.

When I was in college back in the eighties, I traveled to Greece. I had an amazing time, but I remember the ridiculous pain it was to call home and say hello. I had to find the phone, get the phone line out, change money, and pay an exorbitant price; it was ridiculous. Now I can FaceTime with my family from the Acropolis, or live stream on Facebook to thousands of people, hosting a history lesson and teaching the world. The opportunities are endless, fascinating, and valuable, for both creators and consumers.

CHAPTER 12

APPS FOR EDUCATION

As a former teacher, I'm most passionate about apps for education, where the opportunities can be transformative. I wouldn't have been able to do half of the things I did in my classroom without iPads, iPhones, and laptops. When you add apps into the equation, we get a broad range of abilities like that of a Swiss Army knife, despite limited budgets and resources. I'll keep my frustrations with limited educational budgets for another book, but the reality is that teachers are strapped, and technology alleviates that problem and gives students the education they need.

By purchasing a few devices, teachers can teach sports, theatre, journalism, filmmaking, animation, photography, and so much more. You used to need tons of money, space, and devices for those things, but now you only need one device and a few apps, many of which are free. It really used to be the dark ages. Now, it's a golden age—no hours needed to develop film, no years of learning to draw for animation. It's the best time in history to be an educator or student.

The students are never bored. They love portable devices—touching apps, moving things around, and making things. Technology is cool, interesting, and educational. It's nothing like opening up a big, boring textbook and poring through pages. It's hands-on and effective.

One great app is Green Screen, created by Do Ink. As the name suggests, it's a simple app that keys out the green of a green screen and allows you to put photos or backdrops in its place. You can even use a screen that isn't green, if it's solid color, but an actual green screen is recommended. My son recently did a book report on George Washington, a boring summary on a piece of paper. With Green Screen, he could have told a "living book report" story, with pictures of George Washington or Valley Forge behind him. An elementary-school kid could have told an interesting story with graphics, had more fun, and therefore been more apt to learn. Green Screen costs $2.99 and allows you to share your creations with the world.

Green Screen, like most apps, doesn't require users to be technology experts. If you're an older English teacher who isn't a tech buff like me, you can still press a few buttons, follow directions, and teach your kids. The app connects to Google Drive and Dropbox, so you can store photos and videos in the cloud and play them back later. If you have an Apple TV in your room, you can stream all of your students' creations on your projector. They can get instant feedback. Green Screen has tons of applications, but the most common has been "living book reports."

When I grew up, visual learning wasn't acknowledged. You were forced to learn from big textbooks, even though we're all built differently. My eight-year-old doesn't watch TV, but he watches YouTube tutorials that teach him how to play videogames like Minecraft, as well as YouTube videos on subjects he learns in school. He's learning at a rapid pace, despite his distaste for textbooks.

Another great app is iStopMotion, which can teach anyone how to learn stop-motion animation. It teaches you *onion skinning*, which is what professional animators used to do, by hand, with tracing paper. Now you can do it with an app. Onion skinning is when you shoot a still of whatever you're animating; let's say it's a Lego. When you move the Lego two inches forward, the app shows you the first shot faintly in the background, so you know where to put the next one. iStopMotion gives you the skill of a professional animator, on demand.

It's so easy that kindergarteners can use it. Teach them how to use Legos or Skittles, and they'll learn how things move in relationship to time. It's advanced science, but they'll listen and understand. They can see it—the Skittle jumps across the screen from frame one to two at first, but when you fix it, it goes nice and slow.

Another app I recommend is Bombing Brain's Teleprompt+. It turns your iPad into a teleprompter, and it films while you read the script. It's customizable and has remote controls, so it's perfect for shooting a news segment or reading a monologue. You don't need anyone there to help you.

It's perfect because you can send students out into the world alone to learn broadcast journalism. They can experiment and create professional-looking content without needing help. They also get practice recording and editing on their iOS device, and they can use their computer for more advanced editing later. It provides complete, self-directed filmmaking and journalism class. They can send their footage digitally to their teacher for feedback, and even submit it to CNN iReport. The workflow is faster, so they learn faster.

Some of my favorite apps are for sports. If you haven't gotten the memo, high-school sports are *huge*. I didn't realize *how* huge until I was put in charge of the San Marino High School field-production team. During a meeting my first week, the sports boosters cornered me and told me they needed multi-camera shoots, to be put on HD DVDs to sell. Yet they had archaic cameras and wanted me to be ESPN on a five-dollar budget. Fortunately, my iOS devices, Wirecast, and Switcher Studio saved the day.

The Hudl app is great for coaches and athletes, allowing you to record practices and games for analysis. It used to be difficult to make highlight reels to send college recruiting services, but the Hudl server will identify individual athletes automatically. A player or parent can search for clips, make subclips, and create a highlight reel, pain free. Hudl also does replays, and it's great for coaches and athletes to analyze film for game preparation and improvement.

Hudl is so quick that you can film at the top of the stadium using the app with a nice wide-angle shot, send it to the server, and the coach will receive it on the sidelines immediately. He can pull it up on his iPad and review the play, or he can show the team at halftime. Thank the good Lord they didn't have this while I was at San Diego State; I would have gotten yelled at all the time!

Professional coaches would have killed for Hudl years ago, but now Pop Warner coaches have access. Hudl has a yearly subscription fee; check their website for current pricing. They have great customer support and come highly recommended.

I also recommend the Coach's Eye app, which is for deeper, more individualized player analysis. It allows you to focus on one player, getting close up on a golf, tennis, or baseball swing. Coaches can draw on the video for advice on stances or swings and can attach audio commentary. Coaching can be done in person or virtually, and players can review videos, drawing, and commentary on their own, with their phone, tablet, or computer. It's fantastic for helping kids improve in their sport of choice.

My friend, Mike McGillivray, is a former USC punter and top kicking coach. Last year's Rose Bowl-winning kicker for Stanford was one of his students. He loves Coach's Eye, using it to show players what they're doing right and wrong. He doesn't understand how he used to coach without it. He's amazed by the ability to have digital files everyone can access, on demand. It has made him a better coach and produced better talent, not to mention made his job easier.

We discussed live video earlier, but that's obviously

another key application for education.

Finally, there's The Cube, a great app that allows you to live stream sporting events or school plays. It even displays scoreboards for sports, among other features. You could have a school Facebook or Twitter page for Facebook Live and Periscope streams. If Grandma is ninety years old, lives in New York, and can't come to Laura's graduation, someone can turn on her computer or phone, and she can watch her granddaughter graduate, live. She can watch Laura's school play or soccer game, and so can Mom or Dad. It's beautiful.

It's such an exciting time to be an educator. Every day, a new app comes out. We're testing them all of the time, so check our Educational Applications page on **iOgrapher.com** for more information. Feel free to contact me directly (**davesbook@iographer.com**) if you have any recommendations, ideas, or feedback.

PART FIVE

WORKFLOWS

CHAPTER 13

IJOURNALISM

〞〞〞〞〞〞〞〞〞〞〞〞

A few years ago, traditional journalism was still in the dark ages, but things have gotten exponentially easier since. You used to need two people minimum to broadcast, one with a giant camera strapped on his or her shoulder. There were cables everywhere, for power and microphones. It was a hassle, and difficult to maneuver in places with crowds. Then you needed to return to your big, expensive truck with that big, expensive camera, upload the footage with those big, expensive cables to the central editing bay, then hopefully have a signal to send the footage back to the newsroom. It was inefficient, expensive, slow, and didn't always work. That's why you've always had to wait until 11:00 p.m. to watch the nightly news; they couldn't have physically produced it earlier!

Today, news reaches the ends of the Earth in seconds. We get instant, constant updates via Facebook and Twitter, then read endless commentary. It's the perfect setting for iJournalists, mobile journalists who can report stories without needing credentials. All they need is their iOS device and some optional accessories.

People often become iJournalists by accident. In a recent, tragic police shooting in Texas, a guy happened to be across the street and got footage of the murderer. It was the only known footage of the crime. He sent it to CNN, got on the show, and helped catch the criminal. I told my students this all the time, and I'll keep telling you: we're all documentarians, and we document our lives every day with selfies, videos, Instagram posts, and so much more.

Anyone can become a journalist; the playing field has been leveled. It was so much more difficult when I got started. I used to report for several production and post-production magazines and websites, including my own, *Making Central*. I'd post videos, interviews, tools, tricks, and tips. Every year I'd attend the National Association of Broadcasters (NAB) convention in Las Vegas: the Disneyland for video geeks. The process was a nightmare, though. I'd carry my bulky Sony camera, with a ten-foot cable, handheld microphone, gigantic tripod, and big light, just to get a few interviews. Then I'd have to bring it back to the newsroom, connect everything, and edit. It was a struggle to get things out the day of the convention.

Now everything fits in my backpack and can be done instantly. The dark ages are over. The logistics are easy. All that matters is the quality of your story.

Fortunately, the dark ages taught me some valuable lessons. Most beginners don't realize that you need an intro and outro to produce quality journalism. For the intro, I'd say something like, "Hey, guys, it's David Basulto here at the NAB Show in Las Vegas," and add something fun. I'd mention who I was interviewing, like Steve Jones from Adobe, then cut to the interview. It would be a two-camera shoot, because all I had was myself and two cameras. Then I'd film the outro. "Adobe's got some great stuff," I'd say in front of their booth. Then I'd film *B-roll*. B-roll is background video; if you're interviewing someone about today's surf, the B-roll will be clips of big waves playing over the guy talking. Then, I'd go edit. All of that can be done on my iOS devices, on the fly.

An iJournalist's video quality will be as good as a professional's, as long as he or she has an intro, outro, quality interview, and some B-roll. It's simple. The only difference between you and a professional using fancy equipment is that you can make your video more quickly. If you edit on the fly, you have an edge.

It's important to follow professional journalism practices, though. You need to look the part. When a TV channel cuts to the reporter, the reporter establishes who he or she is. "Hi, I'm Ann Miller, with LA Live, where we're going to see Kobe Bryant play his last game." In movies, they do the same thing, with an "establishing shot." It sets the tone of where you are—for instance, on a transport ship, approaching the Death Star, in *Star Wars*. Then, you get to the meat. Don't jump right into the story like most beginners.

The meat might be Ann interviewing fans about Kobe. Then she goes inside and watches the game, and she films some B-roll of him playing. B-roll is essential so we don't have to watch a bunch of "talking heads." Show the people who or what you're talking about.

Then, there's the outro. "We're live once again, at LA Live. That was Kobe's last game, and people are crying out here. Thanks for watching." Those are the basics.

My students used to make the mistake of telling a great story in front of a blank wall. It drove me crazy. We were in San Moreno, California, surrounded by the beautiful

San Gabriel Mountains, but they would shoot in front of the blank wall outside of my classroom. They'd ignore the beautiful flowers, trees, and homes. Eventually, I taught them the importance of background substance, and they improved. If you're talking about Friday's football game, have the field in the background. Finding a good location is essential. There's always somewhere interesting you can shoot.

In the prior chapter, we talked about Teleprompt+, the iOS teleprompter app. Most people can't talk off the cuff and should use a teleprompter. The app is simple; it allows you to frame yourself in relation to the camera, adjust text size and color, and write your script on your laptop or iOS device.

Set up your shot, open the app, change it to fast or slow, and perhaps use their Bluetooth remote to start, stop, and pause. You can practice on your iPhone. A teleprompter used to be an expensive luxury, and now you have one in your pocket. You'll never forget a fact again.

You don't want to be one of those people with your nose embedded in your iPhone, but you don't need to speak impromptu. Most people can't. It's a matter of preference. You can have brief notes to refer to or a full script to read verbatim.

One person can create a great piece of journalism by himself or herself. *Telemundo* once sent a reporter to Mexico City by herself, with nothing but iOgrapher equipment and the Teleprompt+ app. They didn't need to send a cameraman and editor. She uploaded the video immediately to their servers, and they started editing back in Florida. It was on TV far quicker than it would have been in the dark ages.

iMovie is great for iJournalism. You can film B-roll and add audio later, talking into the camera and describing what you see, as you play it back. It's incredibly quick for simple edits, and it's literally a drag-and-drop process. You can slow down footage or speed it up, depending on how much you have to say. The only weakness is you can't yet add B-roll *with* audio; you have to add the audio separately. People are working on solving this, though. For now, go into iMovie, do your rough edit, and bring it into Final Cut Pro to add B-roll.

National Geographic has a fix for the B-roll audio problem. For example, say they're interviewing a guy in the Amazon, talking about piranhas. They film the interview, then go film the piranhas. Afterward, they record the same guy speaking over the piranha footage, in iMovie. Adding audio later is easy.

Stephen Quinn, coauthor of *MOJO: The Mobile Journalism Handbook: How to Make Broadcast Videos with an iPhone or iPad*

For me, the key issues are autonomy and speed. I can produce a news video entirely on my iOS device doing everything myself: planning, shooting, editing, narrating, adding captions, and uploading. I remember when I was a TV journalist how frustrating it was having to tell the editor what I wanted, or writing instructions on a piece of paper or email if I was elsewhere. Now I do it myself, and I am much faster and happier as a result.

Go to a news event. Shoot video and stills. Interview people on location. Find a coffee shop and order a large white coffee. Edit on location with iMovie. Find a quiet location and record narration directly into iMovie. Add name supers, a headline, and credits. Then I use my portable Wi-Fi device to upload to the newsroom or YouTube. Then I phone the news desk to make sure they are happy, and move on to next story. Or I stay in the area to send updates, depending on negotiations with the news desk. For feature pieces, the process is the same, except I might compose my script and do narration in my hotel instead of a coffee shop. See my book for more details.

Here are the pros of mobile journalism: speed, autonomy (I do everything myself and do not need to worry about others slowing me down), a chance to get cool video from angles that a standard video camera cannot find, a chance to be on location and producing quickly, and there's less chance of authorities and border officials confiscating my iPhone than a big camera.

Here are the cons of mobile journalism: I need to focus on sound (good mics vital) when outdoors in wind, need to ensure backup for battery (iOS devices gobble battery life), need to be alert for people stealing equipment, and need to use tripod and body to ensure quality images.

Jennifer Matthews, Turner Broadcasting

Live video is a quick way for me to gather video, audio, and text when breaking news as well as gathering sources.

In Louisiana during the floods...it was way easier stuffing my "mojo" (mobile journalism) kit in my pocket and using it when needed as opposed to lugging our bigger cameras around to catch the actions.

We don't have a zoom that is reliable on the iPad or iPhone, and changing the lenses on the stabilizers can take up unnecessary time, but it is an easy kit to put together to cover stories on the go and easy to send content back to newsrooms.

As discussed earlier, you can use Adobe Premiere Clip as an alternative to iMovie. The biggest difference is that after you do your rough edit, you can transfer the footage via Wi-Fi to the Adobe Creative Cloud. If you have coworkers in a studio, they can access the footage instantly (if they have access to your account) to do a more comprehensive edit. You could be in New York filming at Times Square, getting great interviews and an outro and an intro, and your main editor in California can download and edit the videos instantly. He or she can put the finishing touches on and share the completed video to the world while you travel home or celebrate with a drink. iMovie is great for iJournalists, but Adobe is better for professional editing, because of their suite of tools.

My favorite thing about iMovie is its various themed template options, especially the CNN iReport template. It displays the CNN logo, so when you film your story, you'll get lower-third identifiers and titles with the CNN brand, as if you worked for the network. You upload your video directly from your iOS device to CNN's iReport page, type in basic information, and you're in the lottery. They have employees looking for keywords and hashtags, and if they like your story, you could make the front page of iReport. I constantly get emails from people who made it, and I love to see the playing field leveled.

The professionals are taking notice, too. We get constant Tweets and emails from news organizations, raving about our products. One of my favorite photos was taken before a news conference during Georgia's Ebola scare, where you see cameras lined up, then all of a sudden, an iPad. Another person sent us a great photo of him preparing to interview Vicente Fox, former president of Mexico, using an iPad. The fact that high-end dignitaries aren't bothered by being filmed by an iPad shows you how far we've come and will go.

Dougal Shaw of BBC, who does video, business, and technology features, loves the iOgrapher's mobility. He recently took his iPhone, a small tripod, a light, and some audio gear around the European Union, interviewing CEOs with ease.

He made a great story, because he's a great journalist, and sent it back to BBC headquarters. They added the identifiers and edited a bit, and it was up on TV.

Nobody waits around to watch the late evening news anymore. Everything is out in the world, in real time. Nobody understands this better than the BBC, one of the most respected news organizations in the world.

I'm enthralled by the possibilities for iJournalism, and

the fact that kids of any age can learn journalism through real-world experience. They just need an iPhone and the desire to tell stories. They can learn the basics of storytelling: intros, outros, B-roll, and interviews. It's simple and free. Failure becomes feedback. You can keep practicing until you become a great broadcaster. There's probably a young girl out there filming football broadcasts, ready to break barriers and become a famous NFL announcer, who wouldn't have gotten the opportunity in the old days.

She can practice in front of her hometown football team, reporting and improving to the point where companies can't help but hire her. The tools are all there. In the past, you had to stare into a microphone and speak into a hairbrush. Nobody would give a kid a chance. Now you can get real-world experience, and see how you did, as if you actually worked for a TV station. You can save twenty dollars from babysitting money, buy the Teleprompt+ app, and tell professional-level stories. You can apply for scholarships and be off to the races.

You don't need to be a tech person to use these tools. One of my dear friends, Jovana Lara, a main reporter for ABC 7 News in Los Angeles, isn't tech savvy. Yet she does regular Facebook Live broadcasts in addition to her main reporting, which allows her to connect with her audience for a change, which is more enjoyable for both sides. It's also great for her personal brand.

An adorable fifteen-year-old girl can visit the Democratic National Convention in Philadelphia, ask presidential candidate Hillary Clinton for a quick, live interview, and she'll have trouble saying no.

As everyone knows, there are two sides to every story. Or if you believe Robert Evans, there are three sides: your side, my side, and the truth. When there are controversial stories like police violence and protesting, you can see all angles, from people of all opinions. It's a wonderful way to find the truth without media bias, and it's a great opportunity for all types of people, of all ages. iJournalism is the future. We can all be reporters.

CHAPTER 14

LIVE VIDEO WORKFLOWS

As I mentioned in chapter 11, the beauty of live video is everyone can do it, and the horror is that everyone can do it. We're bombarded with videos from everyone about everything, because our brains process videos more quickly than words. Live video is what people want; it's easy to consume, and it can be used for anything. You can use it for a political movement. You can use it to market your products. You can use it to watch sporting events. When harnessed correctly, the power of live video is massive. Yet most people don't harness it correctly, because they don't prepare.

Live video's biggest advantage is that it's interactive. When

I had more than four hundred people following me on the battleship *Missouri*, people weren't passively watching. They were asking questions and contributing, and we were having a dialogue. It made it onto Apple TV's Periscope feed, so people were watching it on their TVs at home, while little old me was on vacation in Hawaii, hanging out alone because my kid was bored.

As an educator, you *need* to harness the power of live video. After being asked many times, I've finally committed to doing a weekly live show. It's the perfect opportunity to educate my customers and fans on how to make better mobile videos. It's a great marketing platform for me, and a great, interactive learning platform for them. Our thirty thousand Facebook, sixty thousand Twitter, and twenty thousand Instagram followers can all learn simultaneously. I can change things up every week: teach a new app, tool, or workflow; bring in the maker of the next-generation iPad; or do anything that crosses my mind.

My friend Mari Smith is a live video guru. She came to the United States as a Scottish immigrant with almost nothing in her pockets, and she rose to become the queen of Facebook. She started out on her own, making videos, but she had so much success that Facebook actually hired her to talk about video and the power of Facebook in business and marketing.

Mari is a perfect example of how to use live video effectively. She isn't there to sell her books or a course she's teaching. She hops on a Facebook Live and bombards you with value. That's what it's about. Give people value, and in the long run, they'll trust you and buy your stuff, if it's good. Whenever she talks, whether in person at a conference or online on Facebook Live, everyone listens. Her sole intent is to teach, and I try to do the same. I don't try to sell iOgraphers when I'm on live video. I'm there to teach you how to use what you have already, whether or not you have an iOgrapher. That's also what I'm trying to do in this book.

I met Mari at Social Media Marketing World a few years ago, after hearing about her countless times. Her presentation on the importance of using Facebook to connect with customers blew me away. She argued convincingly that Facebook was a requirement for customer service and education. Ever since, we've worked together on a variety of things. She's incredible and has a great demeanor. Mari loves the iOgrapher and has done giveaways on our show, for which I'm grateful. It's also the perfect example of why live video is so important.

My statistics show that the majority of iOgrapher's traffic comes from Facebook, so Mari has been proven correct. It's funny, because in the old days, you needed newspaper, radio, magazine, and TV ads to get customers. I get emails from companies telling me to advertise on those media, but they sound like dinosaurs. Those days are over. Live video is the newest, best way, and people like Mari Smith get it.

We're in a trust-relationship world. If someone like Mari Smith vouches for Dave Basulto on her timeline, her followers are going to check out my stuff. It's all about building your audience's trust. They'll share with their friends, and the cycle is never-ending. We rely on reviews from our peers. That's the way things are these days, and that's why live video is so important to connect with users, fans, friends, teammates—everyone. It's the ultimate trust builder, because it's interactive, immediate, and transparent.

Mari has gotten me fans, because she's a great person, and when a great person vouches for you, her fans will listen. That said, she's only vouched for me because I've built trust with her through good, old-fashioned relationship building.

Her whole tribe sees my stuff, and anyone who's interested will join my tribe, too. An iOgrapher only costs them sixty dollars, and it's no accident that she and her fans are using our stuff for Facebook Live, plus microphones and other accessories. It's good for us, and good for them.

You get it: live video is important. Now, let's go through the nuts and bolts of the workflow. Occasionally, you'll want to do something fun and impromptu, like my battleship *Missouri* broadcast. If you're doing something interesting—maybe you're at Disneyland and they have a new rollercoaster—take your fans along for the ride. When doing deeper stories, though, you'll want to plan beforehand.

Start by coming up with a topic you want to discuss. Then, think: *What types of questions might my viewers have?* Being prepared will make you more engaging and will provide more value to your viewers. My shows are about an hour long. I start with an introduction to the iOgrapher, because I figure some people are new to the channel and don't yet know about it. I keep it brief, though, so my longtime fans aren't annoyed. Then, I dive in to the meat of that day's topic. We end with time for questions and answers and then a giveaway.

Ask yourself: *Who is my audience? What do I want them to get from this live stream? What do I have to offer?* For example, if you're at a conference, play, or sporting event, you will want to display what's going on to people who couldn't attend. Remember: You can use Switcher Studio or Wirecast to provide more than one angle, which will make you look professional and set you apart from the masses. Also, make sure to add your logo when using one of those apps.

You'll learn that and more in the *iOgrapher Academy* training (**bit.ly/iogacademy**).

Using multiple angles can be a big advantage in certain scenarios. One of the most obvious would be a sporting event, but another would be a simple interview. If I was interviewing Mari, and I was on stage with her, I would probably want three angles: one wide-angle shot that showed both of us, one close-up that showed me, and then another close-up for her. I might even want a fourth angle of the audience, so I can film their laughs and other reactions. Then, of course, I can cut from angle to angle—briefly for things like laughs, and longer for when Mari is telling a great story. If you're filming a high-school play, you'll probably want a wide shot of the whole stage, then close-ups on the main actors. It's as if you're a professional TV station, with options galore, despite broadcasting live.

As of this writing, the only apps worth broadcasting from are three power players: Facebook Live, YouTube Live, and Periscope. When deciding which one to use, think of your audience. Which platform do they interact on? If you're already a YouTube star, or have a big Facebook or Twitter page, the answer is easy.

My son loves a nine-year-old kid "named" "evantubehd," who talks about video games and got over a million subscribers in a year. He should stick to YouTube Live, because he has a huge platform. If Evan was at the big E3 Gaming Convention in Los Angeles, he might want to go on YouTube Live to show his conversations with video-game creators.

For someone who is more of a marketer, Facebook might be the better place to go. That's obvious for someone with a large Facebook page or group, like Mari Smith, but it's also for beginners. If you have a private Facebook group or page, Facebook Live is the place to go. I'm no big YouTube star, so I focus on my Facebook page, and I occasionally branch out to Periscope to appeal to my Twitter following.

The logistics are easy. For Periscope, you enter the app, hit play, and your feed goes right into your followers' Twitter feeds. They can watch from their Twitter time-lines. Sometimes, I'll go on Periscope to introduce my show, then tell people to head to Facebook Live if they want to hear the rest, because Facebook is my main focus. Again, it depends on your situation, but Facebook Live has the bigger potential audience and is more visible, so it's usually the best for those who are not YouTube stars or Twitter celebrities.

Periscope may be the best for beginners, though, because on it you can build a following. Strangers will see your broadcasts. Once you do that, you can branch out. That said, I'm sure you have a Facebook page, even if it's your own, so use it to practice! Broadcasting live video to your Facebook friends is another great place to start. Experiment, and see what works.

We mentioned Chewbacca Mom earlier. All she did was hop on her Facebook and stream live to her friends, to show them how excited she was to get this Chewbacca mask. That turned into millions of views, her own doll, and a meeting with J. J. Abrams. Periscope and Facebook Live are both great for people getting started. So, get started.

Michael Stelzner, Social Media Examiner

Which platform do you use for live video? Facebook Live, Other

Why is live video important in your workflow? We use live video to produce a weekly show. It provides great value to our audience and helps us think through what the industry changes mean.

What are your go-to tools for live video? iMac, Wirecast, Huzza.io

Ryan Anderson, Bell Company, Summit.Live

Website: Ryan.Live

Which platform do you use for live video? Facebook Live, Periscope

Why is live video important in your workflow? I run a conference centered around live streaming. It's an important way to connect with the community and amplify what we're doing.

What are your go-to tools for live video? iOgrapher is definitely one of them. When I'm doing professional broadcasting, I like to use a good lens like an Olloclip and a good mic (Rode or MXL).

Mari Smith, Mari Smith International, Inc.

Website: www.marismith.com

Which platform do you use for live video? Facebook Live

Why is live video important in your workflow? Live video creates more intimacy with my audience. It helps my fans and followers know that I am approachable, personable, real, authentic, helpful, and knowledgeable about my subject.

I really enjoy acknowledging my "peeps," as I call them. People love to hear their name and know that they are important and matter to me. People also love my bubbly personality and Scottish-Canadian-Californian accent, which is not conveyed in the written word!

I've always said that there is no sophisticated technology in the world that will ever take the place of real, in-person connecting, no matter how advanced VR/AR and AI get. When in person, you can shake someone's hand, look into their eyes, read their body language, and feel their energy and their "vibe."

The next best thing to in-person is video—especially live video. With my niche being Facebook marketing, it's a fast-moving platform in a fast-moving industry. I have to stay up to date, and live video offers a terrific opportunity to keep my community informed with the latest news and important updates.

What are your go-to tools for live video? For simple, quick, and easy, I broadcast on Facebook Live using the Facebook Pages Manager app (or sometimes the Facebook Mentions app) on my iPhone 6 Plus, housed in an iOgrapher case, with a small tripod and iRig lavalier mic. This setup is perfect for on-the-go mobile broadcasting anywhere, anytime. I use it at my home office, out in my back yard, or on the streets of Rome or Edinburgh! I make good use of the front- and back-facing cameras to give my audience a good feel for where I am. And/or, I focus the camera on my computer screen if I'm sharing something online. When I want to get more advanced, I broadcast on Facebook Live, using Wirecast on my iMac desktop.

I use my Logitech HD webcam and my Heil microphone on a boom, shock mount, and popless filter with Shure USB mixer. With high-speed Internet, this ensures the best-quality audio and broadcast. I use a combination of webcam, desktop sharing my browser, and prepared slides.

I've used Open Broadcaster Software prior to investing in Wirecast, and now vastly prefer Wirecast. I recently got an ALLie Camera and battery pack for broadcasting live in 360 degrees on YouTube. At some point soon, I'm sure Facebook will also have the ability to broadcast live in 360, and I'll be using my ALLie Camera for that, too!

CHAPTER 15

FILMMAKING

||||||||||||||||||||||||||||

When I was teaching, I preached to my students that they had everything they needed to become storytellers. Go out and tell stories, I'd insist. I was a kid that loved to write down little plays, too, but that was where it stopped. The barriers were too big. The prices of cameras and film, the developing film and editing, the projector to show it on, the price and risk of going to film school—it was too much. My storytelling abilities as a kid were thwarted, but that isn't the case for kids with iPhones in their pockets. The world is their film set, and their studio is in their pockets or backpacks.

Every semester, I'd ask: "Which one of you is going to be the next Steven Spielberg? Who is going to rise to the top and be the greatest storyteller in the room?" I'd dangle little prizes, and encourage them to make things. It was remarkable how much they progressed.

I'd tell them: "You have the same tools professionals use— the same tools that Bentley used to make a commercial, and the same tools used to make *Tangerine*, which won the Sundance Film Festival and got a distribution deal. All you need is that iPhone or iPad. The playing field is level. There's no barrier. It's up to *you* whether or not you'll be the next Steven Spielberg." I love the empowerment and meritocracy, and so did the kids.

To start, I had them make what I called a "war chest." That's simply an inventory of what you have access to when you're starting out and may have a zero-dollar budget. Even beginners have *something*, whether it's a cabin in the mountains, your mom's car, ten friends who can help you, old blood makeup from Halloween, or a nice park nearby. Write down *everything* you can get your hands on, so you know your boundaries, and then build a story and a script around that.

I made my own war chest once I left "big Hollywood" and no longer had resources at my disposal. I wanted to make my own film (though this was pre-iOS devices), called *Death Clique*. It was about ten little Native American friends who were dying off one by one, but nobody could find the murderer. It had fascinated me as a child, and scared the hell out of me. I wanted to tell the story through film, so I made my list. I had access to my parents' house up in the hills of South Pasadena.

I was still friends with people at my hometown's high school, so I could film there and have a school setting. I could drive up to the local mountains for free. I didn't need a permit for any of that. We put together *Death Clique* for under $10,000, it was sold to Blockbuster, and it's available on Amazon today. That was twelve years ago. Nowadays, I could have made it quicker, for far cheaper, by using iOS devices. For $10,000, it could have been an epic movie.

While screenwriting is not my forte, I can offer tips to beginners. Start with Sid Field's book *Screenplay: The Foundations of Screenwriting*. It gives you a step-by-step guide to go from concept to delivery. I read it to give myself an idea of script quality when I was flooded with scripts as a Hollywood film producer. After reading that book, then tons of good and bad screenplays, I have a good eye for what makes a good script. Everything pointed back to what Field had written, so I highly recommend starting with that book.

Next, make sure you read scripts. You can buy the scripts for your favorite movies online; that's the best way of understanding how they were written. You'll get a sense of their tempo, which will be essential. I personally have never written a script, and don't plan to, but reading that book helped me recognize and appreciate the genius of writers I've worked with.

We hired a budding writer for *Death Clique*, but my understanding of screenwriting allowed me to communicate my idea to him, show him my "war chest," and then let the genius get to work.

Once your script has been written, it's important to practice shooting film. You need to understand the different types of shots, so you can create the flow your movie needs. I used to give my students an assignment I called "The Shooting Gallery." I got it from a friend I played college football with, John Corippo. He created a PowerPoint deck with all the basic shots in every single movie, based on his own research. It was comprehensive and powerful.

I made my students film an example of each shot: close-up shots, extreme close-up shots, wide-angle shots, voyeuristic shots, two-shots, over-the-shoulder shots, and so on. Then I'd have them watch a bunch of movies, and explain at which points each shot was used. That taught them the value and specific use of each shot, which unlocked their storytelling abilities.

With your iOS device and iOgrapher, all you have to do is download "The Shooting Gallery" template (bit.ly/shootngallery), get out, and practice. You don't need anybody with you. You could film action figures. There's nothing stopping you. Go do it.

Once you've written a script and learned your shots, you need to create a storyboard. How are you going to relay your story to everyone working on your video/movie? Believe it or not, George Lucas originally told *Star Wars* on a storyboard that he drew himself, with stick figures. He had all of the resources in the world, but he was able to communicate the Battle of Endor, and every other *Star Wars* scene, with crudely drawn mountains, landscapes, and stick figures. If you look at them today, you'll recognize everything and understand his point, like his production crew did.

I personally use an app called Paper, because I'm not a good artist. I used it to create my first iOgrapher prototype drawing; I needed to get the idea out of my head. It's also great for storyboarding. You can use a pen or stylus, but I use my finger. You don't need any artistic ability, only a point that you want to get across. If the great George

Lucas can express his creativity through stick figures, so can you. You have no excuse.

Storyboarding is essential, especially if you're a director or writer who isn't doing the camera work. The cameraman needs a storyboard to know what to shoot, and the actors need one to know where to be and how to act. It's a clear, visual example of the story you're trying to tell, so everyone can get on the same page and carry out your vision. If you were a general sending your troops into battle, you'd send your soldiers a war plan. They need to know how the battle is going to flow, so they can understand and implement your plan. The same goes for filmmaking.

To reiterate, there are no barriers to the videos you can make with your iOS devices. You have all of the tools you need for preparation, filming, and postproduction. One of my favorite examples is Richard Lackey, who used DaVinci Resolve, a free color-correcting tool that used to cost $200,000, to touch up iPhone footage shot in Chicago. The footage looked professionally filmic. You would have had no idea it was shot on an iPhone. The article is a must-read, and you can find it at **bit.ly/LACPUGIP**.

You have every tool you need at your disposal. All you need to do is come up with a story, put your resources together, and get to work. You don't need permits. You can walk around your city, get some friends together, and send your film into Sundance, like *Tangerine*. You can get on iTunes, get a distribution deal, and have screenings across the country. If there's a will and a story, there's a way.

The naysayers might say you need fancy cameras like the Red Camera, or the ARRI Alexa. Maybe one day you'll be filming the next *Lord of the Rings* (shot on the Red Camera), so you'll want one then. For now, you have everything you need in your pocket. The only way to move up the ranks is to start making things.

Realize that the only thing stopping you is you, and get out there and tell stories.

CHAPTER 16

EDUCATION

After creating the iOgrapher, I was still a teacher, and I needed a way that fit within a school budget to film class projects and sports. During football season, most cameras were checked out for the weekends, so students couldn't use them. iOS devices were the solution, and I haven't looked back since. I started in my classroom, but word spread to other teachers quickly.

First, the English Department reached out. The teachers didn't know much about technology, but they saw the potential to film their Shakespeare events, and knew how to hit the camera button on an iPad. It took almost no effort to teach them how to set up and add a simple microphone. We were fortunate to have Apple TVs in our classrooms, so they'd film the plays, and students could watch them easily in class.

The play could be streamed right from the iPad, the following day, so that students could critique and analyze the play, themselves, and each other. That immediate, visual feedback showed great results—far better than writing a paper, putting red marks on it, and having students throw it away. The students saw how they did, wanted another chance, and improved the second time around.

Media teachers should brainstorm ways to work with other departments. Don't simply lend out your devices; work together. I worked with Spanish classes, where students would write commercials in Spanish, then film them in my class. It was a great way for students to practice both writing Spanish and visual storytelling, and they wound up with an awesome product in the end. By using each department's strengths, you can maximize results. Your school's drama department is full of actors; use them! It's the perfect chance for them to practice acting, and for other students to practice filmmaking. Actors can work with anyone, whether it's media classes, journalism classes, English classes, or even science classes. Reach out to other teachers and students for ideas, and let your imagination run wild.

As I've mentioned before, the best way to make your school's booster club happy is to create professional-quality DVDs of sporting events, especially football. By using iPhones, iPads, and the iOgrapher, that can be done cheaply and easily. No more big, old, expensive broadcast cameras with tapes in them. No more waiting days to develop film, then more to edit and render them on DVDs.

By using iOS devices and iOgrapher cases, you save tons of time. As mentioned earlier, we'll put one camera on top of the stadium, shooting wide angle, then have two close-ups on the ground. Everything is shot in high definition. We either take the footage back to the classroom to edit, or we send the iPads home with the students so they can edit there. It's far quicker than the old way, and if you have Wi-Fi and use an app like Wirecast or Switcher Studio, you can have your edit done *by the end of the game!*

Once it's ready to go, you can share your video on a site like Vimeo and sell it to interested viewers. If you tell the booster club that Sunday's game will be available for purchase on Monday, they'll be thrilled. Parents can download their kids' games to cut up for recruiters, or for film study for the kids' improvement. The workflow is quick and efficient.

To film a football game, you'll want two angles, minimum. The more angles, the merrier. A wide-angle shot from up high is your first priority. Next, you'll want a fluid-head tripod, which allows you to pan in all directions, smoothly. You can find one on Amazon or iOgrapher.com—check chapter 7 for recommendations. Then, if you can, add a telephoto or close-up shot, possibly using our 2x-telephoto lens discussed in chapter 9.

Next, you'll likely want to capture audio from the game's announcer. The two angles will pick up audio from the stadium's loudspeakers, but that will probably be of poor quality. I would recommend plugging in a microphone right in front of the announcer. The cheap option would be to use a small Lavalier mic, plugged into another iPad running an audio app like Apogee MetaRecorder, which allows you to sync the audio in postproduction. For a more professional option, you could use the Apogee Duet, Apogee Jam, or the iRig 3. If you have one of those, you can buy a twelve-dollar gaming headset on Amazon for the announcer, and run an XLR cable down to him. That will give you ESPN-like audio. We did that recently, with the iRig 3, and were able to get the announcer's audio directly to one of our iPads, so we were already synced for postproduction.

It's also important that you use high-quality settings. If you're using regular iPad cameras, go into settings and ensure that 1080p is selected. If you're using the new iPad Pro, you could shoot in 4K, but the file size for a long football game will be massive. I'd recommend 1080p. Then, ensure you're shooting at thirty frames per second. Some apps use twenty-four frames per second, but that's not enough. Twenty-four frames per second will give you more of a filmic look and is better for movies.

When you're done filming, ideally you'll upload the footage via Wi-Fi and Wirecast or Switcher Studio. If you don't have Wi-Fi, you'll have to wait. Once you've uploaded the footage, you can bring in your two or more angles into Final Cut Pro or Adobe Premiere Pro; both are great, simple tools for multi-camera edits.

As I've explained, iOS devices all use the Apple clock, so your shots are synced automatically. No clapper board is needed. In Final Cut Pro, you select your two clips, right-click, and click "multi-cam edit." The program does everything for you. All you need to do is switch back and forth between the angles, and it'll make cuts automatically. You press "play," click on the piece you want to include, then press the other angle when you want to switch. Final Cut Pro makes the cuts automatically. Premiere Pro does the same exact thing.

If you had audio on one iPad, from the announcer's microphone, you'll use that audio for the entire game. That's easy. After you've done your edits, you can add lower-third titles and a scoreboard.

Once you're done, upload to Vimeo, and charge people if you'd like. The boosters used to want Blu-ray DVDs, but those were a pain in the butt. They often wouldn't sell, and we'd be stuck with them. Now the footage is digital, there's no inventory, and students can access archives years into the future.

As discussed earlier, you can use an app called The Cube to live stream football games. You can add a scoreboard, have people leave comments, and it's a wonderful workflow. Viewers can download a clip for ninety-nine cents—so if Junior breaks a hundred-yard touchdown run, Papa Senior can capture it in real time.

That said, many schools have their own Facebook page, which means they can use Facebook Live. You can stream directly from Facebook's app, or via Switcher Studio if you want to add up to four camera angles, plus graphics such as pictures of the quarterback. You could do a quick cut to the quarterback's picture after he makes a nice play, and tell a quick story. I highly recommend using Switcher Studio and Facebook Live in conjunction with each other for sports streaming. YouTube Live is another good option.

The most robust tool would be Wirecast 7, which allows you to stream to many options, including The Cube. You can shoot multiple angles, add scoreboards in real time, and go as big as you want. It's a professional-level tool available on your ordinary laptop.

One of my favorite examples of live streaming is high-school graduation. Your British grandmother could watch you graduate at a California high school while she rides the Tube in London. Neither location nor health needs to stop family and friends any longer from watching events live. The level of connection we have is amazing.

iOgrapher was born because my students inspired me. If you're an educator or parent, let your kids inspire you. They're full of vitality and creativity. Collaborate with other parents and teachers. Take advantage of the endless opportunities at your disposal, and let the best stories get told.

HUMAN STORYTELLING INSTINCT

Even in the cavemen days, humans were compelled to tell stories. It's what we do. If you look at famous kill sites, or places where anything happened at the dawn of humanity, humans wanted to tell stories visually. There were crude paintings, and then there were eventually books. It all has led up to this moment in time, when a device that fits in our pockets can tell a visual story to billions of people.

Social media allows things to spread quickly, and if your video goes viral, millions of people will see it. You can tell stories to the masses, to influence and entertain. It's the best time in human history to be a content maker and storyteller, with everything you need in your pocket.

My nine-year-old son rarely watches television. All he wants is to be entertained and learn, and he finds that on YouTube. Whether it's silly, funny videos from random kids, serious videos from storytellers, or video-game explanations from his favorite makers, he doesn't discriminate. All he cares about is entertainment quality. He wants to watch whatever he wants, on demand. He doesn't care if something was filmed with an $80,000 Red Camera in a Hollywood studio.

His generation wants to be entertained. That's why Snapchat has over one hundred million active users. That's why Instagram has over four hundred million active users and was sold to Facebook for two billion dollars. It doesn't matter which tools you use, or who you are. All that matters is that your story is entertaining. Go out and tell stories. Don't worry where you are now. The opportunities to practice are endless, and you'll get better. The world is waiting to see which stories you have to tell.

One of my son's favorite YouTube channels to watch is by a random guy in London, Dan TDM (The Diamond Mine Cart). He talks about different video games, whether it's Minecraft or games about robots, and he is a general entertainer. He has over twelve million subscribers. He did a Minecraft video two days before I wrote this, and it already has 1.6 million views. He's been on YouTube for four or five years, but he's a young guy, and he's making a king's ransom. He didn't have to beg TV networks to listen to him, and they would have laughed anyway. He didn't need to convince anyone of his worth. He filmed himself in his bedroom, and now he makes well over a million dollars a year playing Minecraft and other video games.

One of my favorites is Casey Neistat, who currently has 3.7 million subscribers. Funnily enough, that's far less than the video-game guy, which shows you the power of video games. Casey is a top-notch, professional film-maker who vlogs his daily life in New York City. He was at the Oscars this year, and he makes incredible content. Recently, he put up a video in which his air conditioner broke. He showed himself walking around New York City for ten minutes, cooling off. The way he tells the story is fascinating, and people are hooked. Find your niche, and tell your story. It can be anything. I keep saying this, because I mean it. Your options are *literally* endless.

If my son was able to make animated videos on his iPhone when he was *four years old*, you can make any type of video you want, with nothing more than that little rectangle in your pocket or backpack and this handy guide. Go tell some stories, and make magic. You already own a portable Hollywood studio, and the world is your set. It would be a shame to let that go to waste; wouldn't it?

JOIN THE
CONVERSATION

‖‖‖‖‖‖‖‖‖‖‖‖‖‖‖‖‖‖‖‖‖‖‖‖‖‖‖

*Now that you've finished reading, join the conversation
with David and fellow mobile filmmakers at*

www.davidbasulto.com

ABOUT THE
AUTHOR

Before founding iOgrapher, **DAVID BASULTO** was a college football player, Wall Street stockbroker, Hollywood producer, and high-school media teacher. His Hollywood career started as an extra in *Rocky V*, and he worked with the likes of Jon Voight, Steven Seagal, and Edward Norton and on shows such as *Cheers*, *Mad About You*, and *NewsRadio*. His most rewarding job was teaching, and his students inspired the invention of the first iOgrapher case. Basulto became a first-time entrepreneur at the age of fifty-one, and he now spends his time helping individuals to create Hollywood-quality films on mobile devices, developing new products, and spending time with his wife and nine-year-old son.